AIRAG KUSHKIK PUMPERMILCH
KUMISS TARHANA TAETTE
MAZUM KEFIR YALACTA

These are the colorful names of some of the fermented milk products nations the world over have prepared for generations. *Why* are these foods so widespread, so popular? Is it true that they can prevent disease and increase vitality and longevity? Or are they, as some have claimed, nutritionally valueless, even in some cases dangerous? Beatrice Trum Hunter has gathered the facts *you* need to know about these foods and presented them clearly and excitingly in her *Fact/Book on Yogurt, Kefir and Other Milk Cultures.*

Beatrice Trum Hunter's

FACT/BOOK ON

YOGURT KEFIR & Other Milk Cultures

Keats Publishing, Inc. New Canaan, Connecticut

Fermentations other than milk products are discussed in *Fact/Book on Fermented Foods and Beverages: An Old Tradition,* Keats Publishing, Inc.

FACT/BOOK on Yogurt, Kefir & Other Milk Cultures

Pivot Original Health Edition published March, 1973

A portion of this book, on lactose-intolerance, is adapted from the author's article, "Milk is Not for Everyone," published in *Consumer Bulletin Annual,* 1973, and reproduced with the kind permission of Consumers' Research, Inc., Washington, New Jersey.

Printed in the United States of America

Library of Congress Catalog Card Number: 72-87858

PIVOT ORIGINAL HEALTH BOOKS are published by Keats Publishing, Inc., 212 Elm Street, New Canaan, Connecticut 06840

To Peggy,
who *thought* she didn't like yogurt

Contents

Why Use Fermented Milk Products?

FOR MY FRIEND, Peggy, to whom this book is dedicated, and for countless persons like her, the sour taste in fermented milk products is loathsome—or so they think. In some instances, these prejudices may simply be from lack of exposure. In other cases, perhaps the food or beverage was pressed upon them by an overly enthusiastic friend, or an overanxious mother. "Try it, you'll like it." "Eat it, it's good for you." Or, the fermented milk product was very tart—which it need not be.

"Can you tell me how to cultivate a liking for buttermilk?" someone asked the late consulting nutritionist, Dr. Michael J. Walsh. Dr. Walsh replied wisely: "The same procedure can be used for learning to like any food. First of all, bear in mind that no human being has yet been born with a liking for, or a distaste for any food, thing, or person. The like or the dislike has been acquired by a conditioning process. In other words, other than a relatively few unconditional reflexes, which include salivation at the taste of food, everything we know we have learned—so here is the conditioning mechanism for learning to like buttermilk."(1)

This was what Dr. Walsh suggested: "Make a point to take one ounce (one jigger) and only one ounce of buttermilk *once* a day, every day, for one week. During the second week, take one ounce *twice* a day, every day. During the third week, take one ounce *three* times a day, every day. During the fourth week, take two ounces three times a day, every day. By the sixth week, when you are taking four ounces of buttermilk three times a day, you have not only learned to like it, but you are likely to be licking your chops in relish as well as having already experienced the beneficial values of it. You understand now why I have no concern for those individuals who insist that they do not like this, they do not like that, they have a natural craving for sweets. Let me repeat—you can learn to like any food, or any person, if you choose to. Similarly, you can readily wean yourself away from foods like sweets, which you know are not good for you in the sense that they do not contribute to your well-being."(2)

What Dr. Walsh suggested regarding buttermilk, can be applied with equal success to yogurt, kefir and other fermented milk products. Are you listening, Peggy and others?

◆ ◆ ◆

"What are the facts concerning the food and health values of the products to which some writers have ascribed unusual values?" is a question answered in the 1965 U.S. Department of Agriculture Yearbook, *Consumers All*. As an amusing sidelight, the entire discussion appears under a section entitled "Food Quackery."(3)

Yogurt is dealt with first. The answer given is this:

"Yogurt has the same nutritive and caloric values as the milk from which it is made. When made from partly skimmed milk, as is often the case, yogurt is lower in fat, vitamin A value, and calories than when made from whole milk. It is a good source of other nutrients in milk, particularly calcium, riboflavin, and protein."(4)

So far, the answer is acceptable. (This information can be verified by checking the tables on pages 36-37.)

The answer is continued: "Yogurt, like other fermented milks, has a fine curd which may permit it to be digested more quickly than plain milk."(5)

This statement is also acceptable. Studies comparing the rate of gastric digestion of yogurt with that of sweet homogenized milk showed that yogurt was digested more rapidly. With both forms of milk, the rate of digestion was most rapid during the first half hour. After that, the rate slowed down. The retardation became more marked with the sweet milk. It was concluded that the bacteria in the yogurt act upon the milk proteins before the yogurt is eaten. Hence the yogurt milk protein is already partially digested (proteolyzed) before it is acted upon by the gastric juices in the stomach.(6)

Then comes the joker, in the U.S. Department of Agriculture discussion: "Yogurt has no food or health values other than those present in the kind of milk from which it is made."(7) This statement is so fallacious that it will take almost the entire remainder of this chapter to refute. As will be demonstrated, yogurt and other fermented milks *do* develop somewhat different food values than are present in sweet milk. And as will be discussed, the fermented milks also have additional health values: bactericidal powers against pathogens, ability to alleviate many gastrointestinal distresses and other disorders, usefulness in relieving antibiotic-induced effects, and a beneficial role in lactose intolerance. Are you listening, Peggy and others?

◆ ◆ ◆

Some food values *are* changed when milk is fermented. For example, during the culturing of yogurt, the biological value of the protein increases. The increase is highest in the yogurt made from cow's milk. The proteins in the milk are believed to stimulate hepatic and intestinal secretions. Lactose is converted to lactic acid, which is regarded as a digestive antiseptic. Alcohol and carbonic acid, which also result, are described as "tonic" to the nerves of the intestinal tract.(8)

It has been reported that infants fed kefir have retained more nitrogen, phosphorus, calcium, iron and fat than those infants fed fresh milk.(9)

Fermented milk products can enhance the usefulness of some minerals. Both calcium and phosphorus may be rendered more available for absorption. In one experiment, rats were fed calcium-deficient diets. When those calcium-deficient rats were fed a balanced diet again, the calcium in it was practically without effect on the serum calcium of the rats. But when they were fed yogurt as a source of calcium, it was better absorbed and used. Yogurt in the ration doubled the intestinal absorption of calcium and enabled young rats to maintain their serum calcium at normal levels.(10)

Some vitamins, notably vitamin B, may be synthesized by the bacteria in the fermented milk, as well as by the bacteria present in the intestinal tract. *Lactobacillus acidophilus* is capable of synthesizing some vitamin B in the intestine. Niacin can be manufactured within our bodies provided we eat adequate amounts of high-quality protein. Niacin is made from tryptophan, an essential amino acid, by the action of vitamin B_6, pyridoxine. The intestinal flora synthesize pyridoxine, allowing the tryptophan to be converted into niacin.(11)

Microorganisms in kefir, in a powdered dry state, were found to contain from two to seven milligrams per hundred grams of vitamin B_2 (riboflavin), while dry kefir itself, was found to contain from six to eleven milligrams per hundred grams.(12)

In testing milk products in southern Tadzhikistan, it was found that riboflavin increased from winter to summer milk (139 and 239 micrograms respectively). Riboflavin increased dramatically in fermented milk products; it rose in *Chekka* (curd) to 625, and in *Kurut* (salted, dried *chekka*) to 812.(13)

Yogurt may show an increase in folic and folinic acids. A Spanish report noted that the amount of these two acids remains high for forty-eight hours after milk is fermented with *L. acidophilus*.(14)

Tests indicate that yogurts vary widely in their capacity to synthesize folic acid. Cultured yogurt is higher in folic acid than yogurt made by a direct acidification technique.(15) (For differences, see discussion of buttermilk processing, page 111-112.)

Researchers in the Soviet Union reported that mare's milk, fermented and made into koumiss, showed a sharp increase in vitamin B_{12}. When koumiss was given to tubercular patients, the increased concentration of vitamin B_{12} in the product was reported to help the patients form blood.(16)

However, the fermenting of milk does not always *increase* food values. According to some test reports on vitamin B_{12} in fermented milk products, the levels are sometimes *reduced*. A Bulgarian report noted that the bacteria within yogurt make use of vitamin B_{12} for their own growth.(17) A French report noted that kefir loses vitamin B_{12} in proportion to the increase of acidity of the kefir.(18)

♦ ♦ ♦

Contrary to the information dispensed by the U.S. Department of Agriculture Yearbook, yogurt and other fermented milk products *do* have unique health values not present in sweet milk. Of primary importance, fermented milk products, in developing millions of powerful lactic acid bacteria, hinder the growth of, or kill outright, some dangerous pathogenic organisms responsible for illness or death in humans and animals. Many of these pathogens, such as ones responsible for dysentery, cannot live and develop in an acid medium such as lactic acid. This powerful bactericidal property of fermented milk products has long been recognized by people in many areas of the world. Doubtless, this property has played a vital role in maintaining health in the absence of sanitation and refrigeration.

In yogurt, *Salmonella typhi* die within thirty to forty-eight hours; *Escherichia coli* are unable to develop; *S. paratyphi* and *Corynebacteriae diphtheriae* lose their pathogenic properties. *Neisseria meningitides* and *Vibrio comma* lose their virulence. Pathogenic or saprophytic bacteria are seldom recovered after being kept from two to four days in yogurt containing from 1.65 to 2 percent lactic acid. Spores do not develop in yogurt unless *Oidium lactis* and other molds have reduced the acidity. It has also been observed that *E. coli, Streptococcus* and *Staphylococcus* were killed in dahi, a yogurt-type fermented milk in India. Although dahi did not kill *S. typhi*, this pathogen was inhibited. When other pathogens, such as *V. comma, S. typhi, S. paradysenteriae*, and *S. dysenteriae* have been added to yogurt, *V. comma* was killed within the first five minutes; *S. typhi*, within an hour; and *Shigella*, within two hours. The acid medium of the yogurt was the deciding factor, since all of the pathogens were able to survive when yogurt was neutralized with caustic soda.(19)

Even month-old yogurt was found to have the same antibacterial effect on pathogens resistant to standard antibiotics. Two researchers evaluated commercial samples of yogurt abroad, and found that both fresh and one-month-old yogurt were effective.(20) (However, it is desirable to eat yogurt not more than about a week old.)

The bactericidal properties of fermented milk products have long made their use popular in tubercular sanitariums. The World Health Organization reported a number of tubercular sanitariums using *airig*, a fermented milk product. Patients receive a liter of *airig* daily in addition to the usual anti-TB drugs. "*Airig* is believed to have a specific effect against TB and it is also claimed that it eases intake of drugs which upset the bacterial balance. The value of the drink *airig* is being investigated by a special research group from the Mongolian Academy of Sciences."(21) The strains of milk yeasts and *L. acidophilus* in kefir produced an antibiotic-like substance which was reported to act bacteriostatically against tuberculosis and typhoid organisms.(22)

The use of fermented milk has become increasingly popular in many quarters for the control of diarrhea in newborn babies and in young children. Fermented milk was reported to be an economical means of preventing diarrhea in newborn infants during an epidemic caused by *E. coli*.(23)

A report from the University of Laval, Quebec, noted that diarrhea developed in *all* newborn babies being fed regular cow's milk; *none*, on fermented milk.

Furthermore, such infections as otitis in the ear, pharyngitis in the throat, or pneumonia occurred less often on fermented than on regular milk. The hospital costs of antibiotics, which had ranged from $358 to $685 monthly, dropped to $79 after fermented milk was introduced. The savings in cost for the antibiotics more than offset the slight additional cost in the food budget with the use of the fermented milk.(24)

Fermented milk, rather than fresh milk, is the preferred food for weaning infants, in the Soviet Union. Pediatricians and nutritionists base their recommendations on extensive research data.(25)

In Sophia, Bulgaria, both healthy and diseased infants were fed sour milk over a period of twelve years. The usual bacteria of the stools were gradually changed and replaced by *L. bulgaricus*, until a pure culture of this bacteria was produced. At the end of the therapy, the ordinary flora of the stools returned, but the temporary displacement of the intestinal bacteria had a beneficial effect in the treatment of diarrhea.(26)

The whey from cow's milk yogurt shows a wide spectrum of bactericidal properties. All pathogenic bacteria tested with this whey, except *Bacillus anthracis*, and *Mycobacterium tuberculosis*, were killed within twenty-four hours; *S. typhi*, *S. paratyphi*, *S. paradysenteriae*, *Brucella abortus*, *V. comma* and *B. subtilis*, within one hour; *S. pullorum*, *S. dysenteriae*, *P. vulgaris* and *M. pyogens*, within two hours; *Br. alcaligenes* and *Ps. pyocyaneus*, within four hours; *E. coli* and *Klebsiella pneumoniae*, within five hours; and *S. lactis*, *C. diphtheriae*, *S. mitis*, *S. fecalis* and *S. hemolyticus*, within twenty-four hours. (27) Fortunately, in many areas of the world, people using yogurt have had the wisdom to drink the whey as well. (For further dis-discussion of whey, see pages 113-115.)

Some of the properties of *L. acidophilus* have been described as "nature's gastrointestinal antibiotics."(28) The writers reported: "In the Near East, despite high incidence of *Shigella* and *Salmonella* dysenteries among visitors in Lebanon, Syria, Egypt and Turkey, the natives seem to be comparatively free of these diseases. It is noted that these people consume leben daily, a composite of curds cultured with *L. acidophilus*." According to the writers, *L. acidophilus* must act to inhibit the growth of dysenteries, food poisoning from *Staphylococcus aureus*, gas-forming yeasts such as *candida albicans*, and other pathogenic micro-organisms. "The two lactobacters, *Acidophilus* and *bulgaricus* appear to act as natural intestinal antibiotics."(29)

A study compiled by *The Lancet* reported that thirteen cases of antibiotic colitis showed good results, and no poor results, after *L. acidophilus* was introduced into the diet.(30) Although each organism survives only an average of thirty minutes, it accomplishes a great deal during that duration to maintain a normal intestinal flora and to ward off invading pathogens.

Of what practical advice is this information? Yogurt can be a good precaution against picking up dysentery when traveling abroad. Onc nutritionist recommended that a prospective traveler, especially one planning to be in areas of the world where dysentery is prevalant, take the precaution of eating a cup to a pint of yogurt daily for several weeks prior to departure. By heavily implanting the valuable bacteria in the intestine ahead of time, the traveler will have increased resistance against pathogenic organisms. Further advice is offered. If no yogurt is expected to be available abroad, or there is any uncertainity about being able to obtain it, the traveler is advised to take along some yogurt in tablet form. Each tablet supplies several million bacteria. Two or three tablets can be taken daily.(31) Even in areas of the world where any fresh milk can be hazardous, or even fatal, yogurt and other fermented milk products are usually safe because of their antibacterial properties. Are you listening, Peggy and others?

◆ ◆ ◆

Another health value found in fermented milk but not in sweet milk is the relief of many gastrointestinal disorders. These include conditions associated with eructation, distension and flatulence. Fermented milks are also reported to improve the appetite, and to stimulate digestion. These cultures have been reported to be helpful in the treatment of such conditions as achylia gastrica, peptic ulcer, cholecystitis, gastroenteritis, colitis, and, of course, diarrhea and dysentery.(32)

The interest in fermented milk for gastrointestinal disorders became much publicized by the work of Ilya Metchnikoff, a Nobelist of 1908, and associated with the Pasteur Institute.(33) Metchnikoff had observed that of all people in Europe, the Bulgarians had the greatest number of long-lived people, including centenarians and older. He studied the diet of these people and found that in addition to their consumption of generous quantities of home-grown vegetables yogurt was one of their great staples.

Metchnikoff examined yogurt, and learned that it contained bacteria capable of converting milk sugar into lactic acid. This acid, he believed, made it impossible for the many disease or toxin-producing bacteria to thrive. As has been discussed above, in this respect Metchnikoff was correct.

However, Metchnikoff had a theory that toxic substances are absorbed into the blood from retained fecal matter and undigested food residues in the intestines. He believed that establishing *L. bulgaricus* in the intestinal tract would replace the putrefactive bacteria, and the supposedly injurious byproducts would not be liberated. He hailed *L. bulgaricus* as "the friendly intestinal flora" and popularized the phrase. He believed that *L. bulgaricus* was responsible for the longevity of the Bulgarians, and that longevity could be extended to all people by establishing *L. bulgaricus* in the intestinal tract.(34)

Following Metchnikoff's pronouncements, all preparation of milk soured with various species of *Lactobacillus* became widely available. Unfortunately, all too soon it was discovered that the effects of these foods on intestinal well-being were only transitory. *L. bulgaricus* is not a normal inhabitant of the intestine, and does not survive long in this environment. *L. acidophilus* and to a lesser extent, *L. bifidus* are the sole types of *Lactobacilli* which are naturally present in the large intestine, and can be established in it.

Researchers working after Metchnikoff demonstrated that as soon as the outside source of *L. bulgaricus* is withdrawn, this bacteria disappears from the stools. However, *L. acidophilus,* a normal inhabitant of the intestine, can easily be implanted in the intestinal tract, especially when administered with lactose or dextrin. Both furnish the best possible conditions for growth of the lactic acid bacteria.(35)

Although Metchnikoff's theory broke down, we should nevertheless recognize his valuable contribution. He promoted the idea that one's body can be protected to a great extent from invading bacteria. This opened up a new field of medical exploration.

We now realize that yogurt in the Bulgarian diet was by no means the entire story. The yogurt was undoubtedly beneficial. But the generous quantities of home-grown vegetables consumed, the way of life, many environmental and hereditary factors, probably played vital roles in the health and longevity of the Bulgarians. Metchnikoff's idea, viewed in terms of present knowledge, would appear simplistic.

Despite the fact that fermented milk products have had a long period of use, even today their health benefits are not well delineated. Authorities agree that the bacteria flora of the intestinal tract are known to exert a strong influence on human health and nutrition. But they are *not* in agreement concerning the effect of nutrition on the intestinal flora.(36) An intake of forty grams daily of lactose, or as much as a kilogram daily of true Bulgarian yogurt "failed to elicit a response in fecal flora."(37) So, it can be seen, that even today, the influence of dietary microflora on the large intestine microflora is unsubstantiated.

Despite these disagreements and lack of complete data, numerous reports have been printed in reputable medical journals, about the usefulness of fermented milk for treating many gastrointestinal and other disorders. For example, two British physicians, Drs. Davis and D'Atto, reported in *The Lancet* that "there is considerable evidence that for some complaints (e.g., gastroenteritis, colitis, constipation, biliary disorders, flatulence, migraine, nervous fatigue) cultured milk can be especially valuable."(38)

Fermented milk cultures have been reported as being helpful with elderly sufferers of chronic constipation. A mixture of prune whip and plain yogurt, prepared and administered to chronically ill elderly institutionalized patients, made it unnecessary for many to resort to laxatives. There were additional benefits: an improved skin tone; and a lessening of seborrheic dermatitis, chronic ileus, and diabetic ulcers.(39)

Yogurt has been widely recognized as a useful substance to control antibiotic-induced gastrointestinal symptoms. When antibiotic treatment has been used, one of the most frequent side effects is the destruction of the normal intestinal flora. This may result in constipation or diarrhea, and either condition produces discomfort. Yogurt has been found beneficial in restoring the normal flora pattern of the intestine. At the same time, the yogurt inhibits undesirable proteolytic organisms.(40) It has been reported that an eight-ounce jar of yogurt has an antibiotic value equivalent to fourteen units of penicillin.(41)

In recent years, the danger of combinations of certain antibiotics has been a cause for concern. The hazards of antibiotic toxicity can be avoided, or their damaging effects corrected, by the therapeutic use of yogurt.(42)

L. acidophilus has displayed effectiveness in reestablishing normal intestinal flora after antibiotic treatment.(43) In addition to the acidophilus milk products available, many preparations are also available in liquid, capsule, and tablet form. Some contain pectin in the form of carob or banana, plus lactic acid and whey; others contain lactose, malt, pectin and gelatin.

Benefits other than in the intestinal tract have been reported for *L. acidophilus*. It has been used successfully in the treatment of an acute infectious inflammation of the gums and mouth (herpetic gingivostomatitis). Eleven patients with acute primary manifestations of this oral disease were treated with a human strain of viable *L. acidophilus* and whole milk products. The results were "sufficiently dramatic to warrant reporting, and indicate the desirability of further, more definite studies."(44)

Among the clinical impressions noted were that "within twelve hours, all patients reported relief of pain; in twenty-four hours, eating without difficulty was restored; and, after seventy-two hours, the patients were lesion-free." An oral dosage of two capsules of *L. acidophilus* preparation had been given four times daily, along with six to eight ounces of whole milk, as a therapeutic measure against this oral disease. In the cases of painful lesions, the capsules were opened and poured directly on the lesions, after which the powder was swallowed along with the milk. The researchers noted that "no side effects and no complications were encountered."(45)

◆ ◆ ◆

Despite the dairy propaganda that proclaims "you never outgrow your need for milk," the fact is that many individuals, especially adults, have some degree of intolerance to sweet milk. They lack a digestive enzyme (lactase) that is necessary to split and utilize the milk sugar (lactose) and digest it. For these persons, a substantial part of the caloric value of the milk is lost, and their ability to utilize other nutrients in the milk is impaired. Undigested and unabsorbed lactose, having strong osmotic attraction for water, acts as a cathartic. This causes diarrhea, abdominal pains and cramps. Undigested lactose may also have unfavorable effects on beneficial bateria that ordinarily flourish in the intestines, with resultant excessive flatulence and bloating.(46) These unpleasant disturbances occur in different degrees of severity, depending upon the degree of intolerance, and the amount of milk that the person has consumed.

At birth, a normal human being is endowed with an ample supply of lactase enzymes necessary to utilize the natural food, breast milk. The lactase level remains high during the normal suckling period of early infancy. After the weaning period, the enzyme activity in the child gradually begins to decline. The rate and extent of this decrease determines whether a lactose intolerance develops. In the adult years, the lactase activity in a normal individual may not be any more than about 10 percent of its original level.(47)

Lactose intolerance exists among a major proportion of the world's adult population. The problem appears to result from dietary, environmental, genetic and possibly other factors. Whereas only 2 to 8 percent of Caucasians and persons of Western European extraction manifest lactose intolerance, the figure is 60 to 90 percent of Greek Cypriots, Arabs, Jews, American Negroes, African Bantus, Japanese, Thais, Formosans, and Philippinos.(48)

It is thought that people who do not consume lactose-containing foods beyond the weaning period have a gradual decline in their lactase-enzyme activity. The world's population may be divided into three groups: (1) those who have never kept dairy animals and who have not used milk in the adult diet (aborigines of Australia, natives of New Guinea and North American Indians); (2) those who have kept dairy animals but who have used milk only in insignificant amounts in adult diets (Chinese, Thais, Philippinos and most African Negroes); and (3) those who have kept dairy animals and who have used lactose-rich foods extensively during their lifetimes (herding cultures of the Middle East, Negro herdsmen of East Africa, and most Europeans and their descendants).(49)

It is believed that, over many centuries, the lack of non-fermented dairy foods in the diet may have resulted in an adaptive decreased activity of the lactase enzyme for those populations found to have limited ability to tolerate lactose. For example, milk consumption has been minimal in Central West Africa, the place of origin for many American Negroes, 70 percent of whom are estimated to be lactose-intolerant. It is thought that the present intolerance may result from early dietary patterns established by their ancestors.(50)

The recognition of lactose intolerance for many individuals may explain, in part, the refusal of some children to drink sweet milk or their having digestive difficulties when they do. The well-established fact of widespread lactose intolerance raises serious questions regarding the soundness of our domestic and foreign milk-feeding programs.(51)

Yogurt, kefir, acidophilus, buttermilk, cheeses, and other lactic-acid fermented milk products *are very low in lactose.* Some nutritionists have suggested that these products should be considered as substitutes for sweet milk in feeding programs.(52) They also suggest another possible alternative, since some people (like my friend Peggy), are unaccustomed to fermented products and have not acquired a taste for them. It is possible to produce milk which is not soured, but differs from ordinary milk by the presence of glucose and galactose, in place of lactose.

Individuals who feel any discomfort after drinking sweet milk should be aware of the lactose intolerance problem. The presence of this condition can be determined by means of a lactose intolerance test, administered by physicians.

Even if milk can be produced which is not soured, but differs from ordinary milk by the presence of glucose and galactose, in place of lactose, other problems may be created by such processing. Galactose creates a problem for some persons, who are born with a galactose intolerance. In infants, this condition, if not promptly recognized, can cause death or lead to mental retardation. (For a discussion of galactose-induced cataracts, see pages 49-52.) Milk containing galactose is also excluded from diets of infants suffering from congenital phenylketonuria, often a cause of severe and incurable mental retardation.

Thus, in rare conditions, unsoured milks containing galactose may be harmful. But, as has been amply demonstrated, fermented milk products offer many benefits, both nutritionally and therapeutically, for most people.

Notes

1. Michael J. Walsh, Sc.D., "Answers to Your Questions on Foods and Nutrition," *Modern Nutrition*, Oct. 1963, p. 8.
2. Ibid.
3. *Consumers All*, USDA Yearbook, 1965, p. 414.
4. Ibid.
5. Ibid.
6. "A Comparison of the Rates of Gastric Digestion of Yogurt and Homogenized Milk," undated reprint of research at a medical institution, supplied by Dannon Milk Products, Inc.
7. *Consumers All, Op. cit.*
8. H. Grenet, Mlle. de Pfeffel, P. Isaac-Georges and A. Wimphen, *Bulletin of the Society of Pediatrics*, 1939, Vol. 37, p. 411; L. Guillemont and Jeremac, *Presse Medicine*, 1939, Vol. 47, p. 295; Harry Seneca, M.D., et al, abstracts of above titles in "Bactericidal Properties of Yogurt," *Amer-*

ican Practitioner and Digest of Treatment, Dec. 1950, Vol. 1, No. 12, pp. 1252-1259.
9. V. K. Tatochenko, *Protein Advisory Group of the United Nations System*, 1972, Vol. 2, p. 34; abstracted in *Dairy Council Digest*, July-Aug. 1972, Vol. 43, No. 4, p. 22.
10. Y. Dupuis, "Fermented Milks and Utilization of Inorganic Constituents," *International Dairy Federation Annual Bulletin*, 1964, Part 3, Vol. 43, pp. 36-37, 39-41.
11. Richard H. Follis, M.D., *Deficiency Disease*. Springfield, Illinois: Charles C. Thomas.
12. M. Schulz and W. Werner, *Zbl. Bakter*, 1942, Vol. 2, No. 105, p. 26; Harry Seneca, M.D. et al, *Op. cit.*
13. V. S. Smorodina, "Riboflavin in milk and sour milk products of southern Tadzhikistan," *Izv. Aka. Nauk. Tedzh. SSR. Otd. Biol. Nauk*, 1962, No. 1, pp. 118-22; abstracted in *Chemical Abstracts*, Sept. 16, 1963, Vol. 59, No. 6.
14. R. Karlin and M. Cartaz, "Vitamin B content of milk fermented by *Lactobacillus acidophilus*," *Syrup Substances Étrangères Aliments*, 6e, Madrid, 1960, pub. 1961, pp. 165-76; abstracted in *Chemical Abstracts*, Aug. 5, 1963, Vol. 59, No. 3.
15. K. P. Reddy and K. M. Shahani, *Journal of Dairy Science*, abstract, 1972, Vol. 55, p. 660; abstracted in *Dairy Council Digest, Op. cit.*
16. R. P. Pulyatov, et al., "Vitamin B_{12} in mare's milk and in koumiss," *Sb. Tr. Uz. Nauchn.-Issled. Inst. Tuberkuleza*, 1961, Vol. 5, pp. 186-88; abstracted in *Chemical Abstracts*, Dec. 9, 1963, Vol. 59, No. 12.
17. *Chemical Abstracts*, 1959, Vol. 53, No. 16, 15397[d].
18. R. Karlin, "Enrichment of kefir in vitamin B_{12} through microbial association of *Propionibacterium shermanii*," *Compt. Rend. Soc. Biol.*, 1961, Vol. 155, pp. 1309-1313; abstracted in *Chemical Abstracts*, 1962, Vol. 56, No. 8, 9171[f].
19. Harry Seneca, M.D. et al, *Op. cit.*
20. A. Yazicioglu and N. Yilmaz, *Milchwissenschaft*, 1966, Vol. 21, p. 87; abstracted in *Dairy Council Digest, Op. cit.*
21. "A WHO Team in Mongolia." *WHO Chronicle*, Feb. 1965, Vol. 19, No. 2, p. 70.

22. V. K. Tatochenko, *Op. cit.*
23. A. Larue, *Journal of Dairy Science*, abstract, 1972, Vol. 34, p. 395; abstract in *Dairy Council Digest, Op. cit.*
24. *Let's Live*, Sept. 1972.
25. Harry Seneca, M.D. et al, *Op. cit.*
26. A. Fykow and J. Mayer, *Nutrition Abstract Reviews*, 1940-41, Vol. 10, p. 162; abstracted by Harry Seneca, M.D. et al, *Op. cit.*
27. Harry Seneca, M.D., et al, *Op. cit.*
28. Arthur H. and Charles A. Bryan, "Nature's Gastrointestinal Antibiotics," *Drug and Cosmetic Industry*, Mar. 1959, pp. 308-309.
29. *Ibid.*
30. *The Lancet*, General Advertiser, Sept. 21, 1957.
31. *Travel-Leisure*, June-July, 1972, p. 27.
32. H. Grenet, et al, *Op. cit.*; L. Guillemont, et al, *Op. cit.*
33. Ilya Metchnikoff, *The Prolongation of Life; Optimistic Studies*, translated by P. Chalmers Mitchell. New York, G. P. Putnam's Sons, 1908.
34. Ilya Metchnikoff, "The utility of lactic microbes with explanation of author's views on longevity," *Century Magazine*, 1909, Vol. 79, p. 53.
35. P. R. Cannon and B. W. McNease, *Journal of Infectious Diseases*, Nov. 1923, Vol. 32, p. 175; abstracted by Harry Seneca, M.D. et al, *Op. cit.*
36. H. Haenel, *American Journal of Clinical Nutrition*, 1970, Vol. 23, p. 1433; D. Paul and L. C. Hoskins, *American Journal of Clinical Nutrition*, 1972, Vol. 25, p. 763, abstracted in *Dairy Council Digest, Op. cit.*
37. H. Haenel, *Op. cit.*
38. *The Lancet*, Feb. 2, 1957.
39. Francis P. Ferrer, M.D. and Linn J. Boyd, M.D., "Effect of Yogurt with Prune Whip on Constipation," *American Journal of Digestive Diseases*, Sept. 1955, Vol. 22, No. 9, pp. 272-273.
40. *Modern Nutrition*, May 1965, p. 5.
41. *Ibid.*, Oct. 1963, p. 9.
42. "Antibiotic Interactions," *Journal of the American Medical Association*, July 10, 1972.

43. Shepard Shapiro, M.D., "Control of Antibiotic-Induced Gastrointestinal Symptoms with Yogurt," *Clinical Medicine*, Feb. 1960, Vol. 7, No. 2.
44. Jerome Lichtenstein, M.D., "*Lactobacillus acidophilus* and herpetic gingivostomatitis," *Journal of Oral Therapeutics and Pharmacology*, 1964, Vol. 1, p. 308.
45. *Ibid.*
46. Fred Kern, M.D. and John Struthers, Jr., M.D., *Journal of the American Medical Association*, Mar. 14, 1966; Dr. Robert D. McCracken, editorial, *Journal of the American Medical Association*, Sept. 28, 1970; Dr. Urs Peter Haemmerli, et al, *Medical News*, Oct. 9, 1964.
47. Dr. Urs Peter Haemmerli, et al, *Disease of the Month*, Year Book Medical Publication, July 1966; abstracted in "Lactose Intolerance," *Dairy Council Digest*, Nov.-Dec. 1971, Vol. 42, No. 6.
48. F. J. Simoons, *American Journal of Digestive Diseases*, 1970, Vol. 15, p. 695; abstracted in *Dairy Council Digest*, *Op. cit.*
49. *Ibid.*
50. *Ibid.*
51. D. M. Paige, et al, *Johns Hopkins Medical Journal*, 1971, Vol. 129, p. 163; abstracted in *Dairy Council Digest*, *Op. cit.*
52. *Dairy Council Digest*, *Op. cit.*

An Old Tradition

FERMENTED milk products have had a long use by many peoples of the world. Although the names given to these products varied, they were esteemed by tradition for their health-giving properties: as preventives against disease, and for their therapeutic qualities.

According to Persian tradition, the method of preparing "yoghourt," "choeneck" or "masslo" was revealed to Abraham by an angel. Such drinks were thought to be responsible for Abraham's fecundity and longevity.(1)

The origin of kefir, also known as "champagne of milk," can be traced back to ancient times. It was referred to as "the Drink of the Prophet." The ferment used for its preparation was called "The Grains of the Prophet Mohammed." Mohammed was credited with the introduction of this fermented drink to his people.

Nearly every village of ancient Asia, Europe and Africa had its own name for yogurt, which was made from the milk of sheep, buffalo, goat, mare, cow, lama, and other animals. The esteem in which fermented milk products was held can be appreciated by the names of these products. They frequently signified Life, Long Life, or Health, as "labben raid" in Egypt; "laban" or "leben" in the Middle East, and "lebeny" in Assyria.

Herodotus, writing about the villagers in the Asiatic Occident, mentioned the fermented milk drink "jasmia" of the Tartars and "ketch" of the Turkestan people.

Lemgo, a thirteenth century historian and traveler through Asia, described a special portion of the Persian Shah's palace, called "Yoghourt-Choeneck," reserved for the preparation of the milk product, "masslo." "Masslo" was similar to "tayer," made by the Jews. It was kept dried and in sacks, transported by camel, and sold in distant villages. Such milk was reputed to prevent epidemics.

Persian women were reported to value the fermented milk product "mosab" to preserve the freshness of their complexions. This belief is still prevalent in modern Iran. The attachment to yogurt throughout the Middle East has always been great. Many a housewife, preparing to emigrate to a foreign land, has been known to spread a cloth with a homegrown culture, dry it, and pack it among her indispensable possessions, to be used as a starter in a new location.

Pliny related that the ancient Assyrians regarded fermented milk as a Divine food. Jenofonte, another ancient historian, described the fermented milk of the Kumaso people, known even up to the present as "kumys." In modern times, the same product is prepared and used in Oriental villages and in sanatoriums, called "kumys sanatoria," especially for tubercular patients. "Kunney" is the product of the inhabitants of the Russian steppes; and of the Mongols, Kalmucks, Kirghiz, and others.

The Tartars in the Crimea also have their kumys, or

"katky" or "koumiss." This is a fermented drink made in southern Russian from mare's milk. It involves the use of yeast and lactic acid bacteria. Romanians also have kumys, and Montenegrins have "kisla varenyka." "Kisselo mleko" is a fermented milk product in the Balkans. Both "kaelder" and "kyael meelk" in Norway appear to be the same as the rather well-known product, "kefir." "Kuban" is a fermented milk made in southern Russia, and characterized by both lactic acid and an alcoholic fermentation.

"Kishk" is a Middle Eastern food prepared by mixing together leben and burghul, a parched wheat, and drying the mixture. The mixture is then ground into a powder and stored for several months.

Similarly, "kushkik" or "kushuk" is used in northern Iraq. It is prepared by fermenting together two parts of yogurt with one part of dried parboiled wholewheat meal. After the mash has fermented for a week, the curd from an equal amount of milk is added. This mixture is allowed to ferment for an additional four to five days. The product is then sun-dried, ground into powder, and stored for a long period of time. Kushkik is so popular in Iraq that the average person consumes as much as twenty to thirty kilograms annually.

The Turkish variation of this food is "tarhana." It is prepared by fermenting two parts of parboiled wheatmeal with one part of yogurt. A variety of vegetables is added, and this mixture is allowed to ferment for several days. Then it is sun dried.(2)

"Taette" or "tette" is a fermented milk product of the Scandinavian countries, in which both lactic acid bacteria and yeasts are involved. The tette, a meadow plant with a blue flower, is basic to its preparation. A few of the leaves are placed at the bottom of a bowl, and boiled milk is poured over them. The mixture is allowed to sit in a warm place, and the milk thickens.

Then the leaves are removed and some of the milk is spooned as a culture into fresh supplies, which formerly in Norway were kept in casks and barrels. The power of the lactic acid bacteria and yeasts is strong enough so that spring milk thus treated is preserved for wintertime use. By winter, it would have changed considerably in character and taste. Great care had to be exercised not to stir it, for the whey separates easily. A little is dipped carefully out of the cask or barrel, and into bowls, to which sour cream may be added.

I, personally, have had a Norwegian taette culture thriving for several years. The cultured product is mild-tasting. For persons who do not like the tartness of yogurt, taette is agreeable. It requires no cooking. After fresh milk is inoculated with a small amount of culture, the mixture is simply stirred and allowed to ferment at room temperature, Once fermented it may be used at once, or refrigerated for future use. The consistency is like that of junket.

The widespread use of fermented milk products is reflected in the various names by which they are known in many places. It is "skyr" in Iceland and the Arctic region; "glumse" in Scandinavia; "taetioc" and "pauira" in Lapland; "langemilk" in Sweden; "ropa," "pimma," and "fiili biima" in Finland; and "mezzoradu" in Switzerland. From ancient times the Dutch called their fermented milk "hangop," a milk similar to the "pumpermilch" of the Germans, the "hocken milk" of the Danes, the "bassmilch" of the Alpine regions, and the "huslanka" and "udra" of the Carpathian Mountains and Bukovina. Other forms of fermented milk products, many of them known from ancient times, include "urgutrik," from Bohemia; "mazum," "mazun," or "matsoon" from Armenia; "gooddu" from Sardinia; "gioddon" from Sardinia, Corsica, and Sicily; "tarho" from the Balkans; "busa" from Turkestan;

"oraka" and "orjan" from Greece, Syria, and Palestine; "mast" from Iran; "dahi," "lassi," "chass," and "matta" from India; and "skuta" from Chile. Yogurt has many variations in spellings and pronunciations including "yalacta" in Norway; "youghourt," "yohurt," "yahourth," "yourt," "jugurt," and "yoart" in Turkey; and "gjaurt yourt" and "yo-urt" from Slavic countries.

"Airig," or "airag," a fermented milk drink from Iran, was described by William O. Douglas as a beverage "with the kick of beer . . . it tickles the nose like ginger ale and tastes like buttermilk and champagne."(3) (For the therapeutic use of airig, see page 11.)

One of the interesting features regarding these various fermented milk products is that many of these peoples have kept their fermented milk strains in pure form as the preparations have been inherited from generation to generation. Many of these products of ancient lineage, esteemed for their health-giving properties, constituted the "secret medicine" of the countries. For example, some of the most celebrated physicians of antiquity recommended the use of fermented milk products to treat dysentery; to treat inflammatory diseases of the stomach, liver and intestines; to strengthen the stomach; to cure diarrhea; to increase appetite; to regulate the heat of the blood and to purify it; and to give a healthy color to the skin, lips and mucous membranes. Fermented milk products were advised "because it is necessary for persons in whom the ordinary milk curdles in the stomach and produces anxiety of feeling to heaviness in the stomach and unconsciousness." Galen, the celebrated Greek physician, noted that fermented milk "is very beneficial for the bilious and burning stomach and it changes the nature by purifying it." He also said that this milk has not the heat and burning quality of ordinary milk for people who cannot tolerate the latter.(4) These observations are timely, as we have new understandings and appreciations of the problem of lactose intolerance.

Notes

1. José M. Rosell, M.D., "Yogurt and Kefir in Their Relation to Health and Therapeutics," Historical Notes, *Canadian Medical Association Journal*, 1932, Vol. 26, pp. 341-345.
2. Clifford W. Hesseltine, "A Millennium of Fungi, Food, and Fermentation." *Mycologia*, Mar.-Apr. 1965, Vol. 57, No. 2, pp. 149-197.
3. William O. Douglas, "Journey to Outer Mongolia," *National Geographic*, Mar. 1962.
4. José M. Rosell, M.D., *Op. cit.*, The Nature of Fermented Milks," *Canadian Medical Association Journal*, 1932, Vol. 26, pp. 341-345.

What Are the Nutrients in Fermented Milk Products?

THE AMOUNTS OF the main nutrients in the milk used in making fermented products are not greatly changed in the process. Table 1, page 36-37, gives the nutrient composition of fermented milk products compared with that of fluid skim and whole sweet milk. The nutritional data should be viewed as approximate, rather than absolute, due to variations in production and raw materials. The nutrient content of most fermented foods basically reflects the nutritional qualities of the milk or cream from which they are made. The medium can be whole, skimmed, concentrated, or reconstituted dried milks that may have been homogenized and/or pasteurized. Cheeses, due to their concentrations, contain increased quantities of most nutrients. Also, microbial fermentation can increase or decrease the nutrient content of cheese and other fermented milk products.

The National Dairy Council admits that there is insufficient published information regarding the nutrient content of cultured dairy foods, especially those other than cheese.(1) Since there are presently no federal standards for the composition of cultured sour cream, cultured buttermilk, cultured milk or yogurt, the nutrient composition can vary widely. For example, the fat content of yogurt can range from 0.8 percent(2) to as high as 20 percent,(3) although generally the concentration of fat is 1.7 to 3.3 percent.(4) Researchers have demonstrated that the biological activity of amino acids and protein differs among yogurts made from cow's milk, sheep's milk, or reconstituted spray-dried milk.(5) As mentioned earlier, during the manufacture of yogurt, the biological value of the protein increases and is highest in the yogurt made from cow's milk. Other researchers, measuring B-complex vitamins, found that four different yogurt cultures varied widely in their capacity to synthesize folic acid. In addition, the amount of folic acid differed, depending on the processing technique.(6) The increase in concentration of B-complex vitamins, except for B_{12}, was dependent on the amount of fermentation time of sour milk, kefir, acidophilus milk, curds, and sour cream.(7)

The nutrient content in cheese is better established. During the curing or fermentation of cheese, a mixed group of microorganisms grows in the milk curd. The major nutrients affected during the curing process are the protein, the fat, and the carbohydrate.(8) Casein, the protein of cheese, is partially hydrolyzed during fermentation and curing. The extent of hydrolysis depends upon the type of cheese. In soft cheeses (Camembert and Limburger) much of the protein is hydrolyzed to water-soluble constituents such as peptides, amino acids, and ammonia. Hard cheeses (Cheddar and Swiss) undergo less extensive protein hydrolysis. The fat in cheese is also hydrolyzed, but not to the same degree as protein. The products of fat hydrolysis contribute to the flavor of the cheese. The carbohydrate, lactose, present in cheese, is converted to lactic acid during the curing.(9) The effect of cultures on the vitamin content of cheese is unclear because of differences in manufacture and ripening processes. One report, disputed by others, is that vitamin A and the B-complex vitamins are stable during the ripening and storage of cheese.(10) Several investigators have shown that the B-complex vitamin content increases during the curing process due to vitamin synthesis.(11) Large variations in fractions of the B-complex vitamin were demonstrated in twenty-three types of cheese analyzed, with differences occurring in vitamin content among samples of the same variety. It was suggested that the highest vitamin concentrations were developed in cheeses due to the biosynthesis of vitamins by microorganisms during the curing process.(12) For example, the microorganisms characteristic of camembert cheese were found to synthesize vitamin B_6. Similar findings were made regarding the effect of the length of ripening of Cheddar, blue and Swiss cheeses. Niacin and vitamin B_6 concentration in blue cheese significantly increased during a six-month period of ripening. The rate of niacin increase was approximately three times that of vitamin B_6. During the first four weeks of ripening, both vitamin B_6 and niacin increased rapidly in Swiss cheese; whereas, during the later phase of ripening the increase was less marked.(13) The vitamin B content of Cheddar cheese was analyzed. During the initial period of ripening, niacin, vitamin B_6, biotin, and folic acid increased; whereas, pantothenic acid and vitamin B_{12} decreased. These two latter nutrients increased, however, during the second phase of ripening. It was suggested that the initial increase in niacin and vitamin B_6 concentration may be related to the lactose content and its subsequent utilization during the curing in Cheddar cheese.(14) Most of the lactose in cheese is

TABLE 1* PROXIMATE COMPOSITION AND MINERAL CONTENT OF CULTURED DAIRY FOODS COMPARED WITH MILK[a]
(amount of nutrient per 100 gm edible portion)

MAJOR CONSTITUENTS

Dairy Food	Food Energy kcal	Protein gm	Fat gm	Total Carbohydrate, gm
I. CULTURED				
i. Cheeses				
Blue Mold	368	21.5	30.5	2.0
Brick	370	22.2	30.5	1.9
Camembert	299	17.5	24.7	1.8
Cheddar	398	25.0	32.2	2.1
Cottage, Creamed	106	13.6	4.2	2.9
Cottage, Uncreamed	86	17.0	0.3	2.7
Limburger	345	21.2	28.0	2.2
Parmesan	393	36.0	26.0	2.9
Swiss	370	27.5	28.0	1.79
ii. Other				
Buttermilk (skim milk)	36	3.6	0.1	5.1
Bulgarian Buttermilk	62	3.2	3.5	4.5
Sour Cream	211	3.0	20.4	4.3
Yogurt (partially skimmed milk)	50	3.4	1.7	5.2
Yogurt (whole milk)	62	3.0	3.4	4.9
11. MILK				
Skim, Fluid (vit. A & D enriched)	36	3.6	0.1	5.1
Whole, Fluid (vit. D enriched)	65	3.5	3.5	4.9

[a]Numbers in parentheses indicate values imputed usually from a similar food. Dashes indicate lack of reliable data for a nutrient believed to be present in measurable amount.

MACROMINERALS

Calcium mg	Iron mg	Magnesium mg	Phosphorus mg	Potassium mg	Sodium mg
315	(0.50)	–	339	–	–
730	(0.90)	–	455	–	–
105	0.50	–	184	111	–
750	1.00	45	478	82	700
94	0.30	6	152	85	229
90	0.40	–	175	72	290
590	0.60	–	393	–	–
1,140	0.40	–	781	149	734
925	0.90	34	563	104	710
121	trace	14	95	140	130
114	0.10	13	88	154	96
102	0.04	9	77	56	40
120	trace	–	94	143	51
111	trace	–	87	132	47
121	0.03	14	95	145	52
118	0.06	13	93	144	50

*Table I and II from *Dairy Council Digest,* Vol. 43, No. 4, July-Aug. 1972, pages 20-21.

TABLE II VITAMIN CONTENT OF CULTURED DAIRY FOODS COMPARED WITH MILK[a]

(amount of nutrient per 100 gm edible portion)

Dairy Food	FAT-SOLUBLE VITAMINS[b]		
	A I.U.	D I.U.	E I.U.
I. CULTURED			
i. Cheeses			
Blue Mold	(1240)	–	–
Brick	(1240)	–	–
Camembert	(1010)	–	–
Cheddar	(1310)	13	1.5
Cottage, Creamed	(170)	–	–
Cottage, Uncreamed	(10)	–	–
Limburger	(1140)	–	–
Parmesan	(1060)	–	–
Swiss	(1140)	–	–
ii. Other			
Buttermilk (skim milk)	Trace	–	Trace
Bulgarian Buttermilk	140	1	0.1
Sour Cream	839	7	0.5
Yogurt (partially skimmed milk)	70	Trace	Trace
Yogurt (whole milk)	140	–	–
II. MILK			
Skim, Fluid (vit. A & D enriched)	200	41	–
Whole, Fluid (vit. D enriched)	140	41	0.13

[a]Numbers in parentheses indicate values imputed usually from a similar food. Zero in parentheses indicate an insufficient amount of nutrient to measure. Dashes denote lack of reliable data for a nutrient believed to be present in measurable amount.
[b]There is inadequate data to include vitamin K.
[c]Reference does not specify whether buttermilk is cultured.
[d]Reference does not specify whether yogurt is made from partially skimmed milk.

WATER-SOLUBLE VITAMINS

C mg	Biotin μg	Folacin μg	Niacin Equiva-lents, mg	Panto-thenic Acid, mg	B_1 μg	B_2 μg	B_6 μg	B_{12} μg
(0)	1.6	36.0	6.08	2.046	30	610	200	1.4
(0)	1.6	20.0	5.15	0.293	—	450	64	1.0
(0)	5.7	62.0	4.78	1.398	40	750	248	1.3
(0)	1.7	10.0	5.78	0.331	30	460	84	1.0
(0)	1.9	23.0	2.48	0.223	30	250	53	1.0
(0)	2.0	43.0	3.08	0.144	30	280	53	—
(0)	2.3	58.0	5.02	1.277	80	500	89	1.0
(0)	1.7	7.0	8.38	—	20	730	84	—
(0)	0.9	6.0	6.35	0.441	10	(400)	84	1.8
1	1.1[c]	11.0[c]	0.75	0.307[c]	40	180	36[c]	0.2
—	—	—	0.68	—	30	160	50	0.5
1	—	—	0.71	—	30	150	—	—
1	1.2[d]	—	0.77	0.313[d]	40	180	46[d]	0.1[d]
1	1.2	—	0.77	0.313	30	160	46	0.1
1	1.6	1.2	0.92	0.370	40	180	42	0.4
1	3.1	6.0	0.92	0.340	30	170	40	0.4

apparently metabolized during the first fourteen days of ripening. Since most cheese is ripened for more than two weeks, it is unlikely that significant amounts of lactose remain by the time the cheese is purchased. In general, the longer the length of ripening, the greater the concentration of vitamins of the B-complex.(15)

Notes

The references for this chapter have been abstracted from *National Council Digest,* July-Aug. 1972, Vol. 43, No. 4, pp. 19-22.

1. "Cultured Dairy Foods." *Ibid.*, p. 21.
2. "Average Composition of Sealtest Foods Products." Research and Development Div., Kraftco Co. 1971.
3. F. Kosikowski, "Cheese and Fermented Milk Foods," Pub. by author, 1966.
4. *Ibid.*
5. J. Rasic, T. Stojsavljevic, and R. Curcic, *Milchwissenschaft,* 1971, Vol. 26, p. 219.
6. K. P. Reddy and K. M. Shahani, *Journal of Dairy Science,* abstract, 1972, Vol. 55, p. 660.
7. L. Gulko and L. Kruglova, Seventeenth International Dairy Congress 1966, EF, p. 689.
8. F. Kosikowski, *Op. cit.;* and V. K. Tatochenko, Protein Advisory Group of the United Nations System, 1972, Vol. 2, p. 34.
9. K. Kosikowski, *Op. cit.*
10. National Dairy Council, *Newer Knowledge of Cheese,* 2nd. ed., 1967.
11. F. Kosikowski, *Op. cit.;* K. M. Shahani, I. L. Hathaway and P. L. Kelly, *Journal of Dairy Science,* 1962, Vol. 45, p. 833; M. E. Gregory, *Journal of Dairy Research,* 1967, Vol. 34, p. 169; K. M. Nilson, J. R. Vakil and K. M. Shahani, *Journal of Dairy Science,* abstract, 1964, Vol. 47,

p. 681; K. M. Nilson, J. R. Vakil and K. M. Shahani, *Journal of Nutrition,* 1965, Vol. 86, p. 86.

12. K. M. Shahani, I. L. Hathaway and P L. Kelley, *Journal of Nutrition,* 1965, Vol. 86, p. 86.

13. K. M. Nilson, J. R. Vakil and K. M. Shahani, *Journal of Dairy Science, Op. cit.;* and K. M. Nilson, et al, *Journal of Nutrition, Op. cit.*

14. K. M. Nilson, et al, *Journal of Nutrition, Op. cit.*

15. F. Kosikowski, *Op. cit.*

Fallacies & Facts

SINCE fermented milk products have been used through the centuries both as therapeutic agents and as foods and beverages, it was inevitable that many claims—some factual, some fanciful—would develop. What are some of the common fallacies? What are some of the little-known facts? It is time to explode the myths, and publicize the truth.

◆ ◆ ◆

Yogurt was advertised by one company as being "known as nature's perfect food that science made better." The Federal Trade Commission judged this to be a misrepresentation, and issued a proposed order directing the company to stop making such a claim. The yogurt company said that an encyclopedia referred to milk as "the most *nearly* perfect food," and therefore it was mere puffing, not illegal, to call yogurt a perfect food. The FTC witness, a trained nutritionist, testified that yogurt lacks certain well-defined nutrients, so that a person who ate nothing else would not be able "to maintain his nutritional status." Therefore, yogurt could hardly be considered a perfect food. The FTC agreed.(1) And so should we.

Commenting on the current yogurt-for-lunch popularity, a New York City nutritionist, Barbara Premo, reported that yogurt is too low in protein to provide a full, balanced meal. It contains only about fifteen grams of protein per cup. Instead it should be thought of as a substitute for milk, contributing calcium to the diet. To make a balanced lunch, Miss Premo suggested that foods from the protein, cereal, and fruit and vegetable groups needed to be added. She suggested yogurt as a dessert, after a sandwich of lean meat, fish or chicken (twenty-five to thirty grams of protein per half cup.) She also advised to eat plain yogurt and fresh fruit, rather than the heavily sweetened fruit yogurts, which are high in sugar.(2)

"Yogurt is a dieter's delight." This statement is fallacious. About 23 percent of all yogurt eaters in the United States select it as a snack or dessert because they are dieting and consider yogurt a low-calorie food.(3) What are the facts? Plain yogurt yields about 152 calories per cup when it is made from whole milk and 124 calories per cup when it is made from partially skimmed milk. This is low, compared to 168 calories for a cup of gelatine dessert, or 294 calories for a cup of ice cream.(4) If yogurt is substituted for sour cream, it has only about one-fourth the calories.(5) *But,* flavored yogurt now accounts for 80 percent of the market.(6) Flavored yogurt was introduced in 1946, with strawberry-flavored yogurt. Since then, many others have been marketed. These fruit-flavored yogurts may contain a very high caloric content, since the fruit flavors are generally achieved by use of a sugar-loaded jam or preserve. The following are some of the caloric contents of flavored commercial yogurt(7), per cup.

apricot from 260 to 286
banana 260
blueberry 260 to 284
boysenberry 260
coffee 200
mandarin about 268
peach about 261
pineapple 286
pineapple-orange 260
prune whip 260 to 284
raspberry 260 to 278
strawberry 260 to 266

The specialty yogurts, such as "Swiss style" and "Swiss Parfaits" are also far higher in calories than plain yogurt. In addition, many of these popular products, sold as yogurt, are a far cry from simple cultured milk. Some, in fact, resemble a cornstarch pudding. When asked "When is yogurt no longer yogurt?" an FDA spokesman said, "We are beginning to take a look. Yogurt may or may not be an entirely appropriate description for a number of such products."(8)

Anyone counting on yogurt as a cholesterol reducer should forget it. Daily consumption of one to two cups of yogurt for six to twenty-two weeks failed to influence serum cholesterol concentrations in nine patients: three normal, and six with unusually high levels of cholesterol.(9)

Anyone counting on fat-free yogurt as a weight reducer needs to be a sceptical shopper. The Connecticut Agricultural Experiment Station, which since 1896 has had a venerable record of examining foods offered for sale in that state, found that certain "99 percent fat-free yogurt" being sold contained a fat content that was high relative to the claim.(10) A similar situation may prevail elsewhere, but not be subject to the same rigorous surveillance as in Connecticut.

◆ ◆ ◆

"Yogurt is a fad food"—fact or fallacy? If we use Webster's definition of "fad" as "the pursuit or interest followed usually widely but briefly and capriciously with exaggerated zeal and devotion," yogurt can hardly be considered a fad. Yogurt, in some form, existed in biblical times, and has been consumed for centuries in many parts of the world. What is true, however, is that yogurt has become popular in the United States only in recent years. According to *Supermarket News*, yogurt sales have jumped 500 percent in the last decade.(11) What is also true is the fact that some yogurt enthusiasts *are* guilty of touting yogurt "with exaggerated zeal and devotion." Here are a few examples:

"Middle Easterners feel that yogurt is the health food supreme . . . [They say] that it confers long life and good looks, prolongs youth and fortifies the soul. Several people also insist it can cure ulcers, relieve sunburn and forestall a hangover. In Iran, girls use it as a facial, and in Iranian villages it is mixed with chopped garlic and swallowed as a remedy for malaria. . . ."

"You'll learn to grow young gracefully with yogurt." This phrase was part of the advertising copy of one yogurt manufacturer. Although the Federal Trade Commission considered this a misrepresentation, the agency was not able to prove that the statement was false. Since the burden of proof is on the government when it questions the truth of advertising claims, the FTC had to drop its objection to this claim.(12)

♦ ♦ ♦

"Yogurt prolongs life." This belief is largely based on Metchnikoff's ideas, and is unsubstantiated. Lack of exact age records, and village pride, have frequently led to exaggerated age claims for many centenarians. A researcher who studied the subject for eight years in villages of the Soviet Union found that the statistics for the numbers of Russian centenarians could be cut from 28,000 to 22,000. Studies of statistics in the United States have cut the number of centenarians from 10,326 down to about 3,700.(13)

At the turn of the twentieth century, probably due to Metchnikoff's influence, many other unsubstantiated claims were made for yogurt. Some claimed that it corrected overweight, restored hair, toned up flaccid muscles and did wonders for the sexual prowess of senior citizens. In more recent years, with the growing popularity of yogurt, we have witnessed more brief but capricious displays of "exaggerated zeal and devotion." A cosmetics company introduced strawberry yogurt-shampoo. A New York City restaurant served its caviar blended with yogurt. A publicity-seeking model ordered 200 pounds of yogurt for a "beauty bath."(14)

♦ ♦ ♦

"There are rare instances when individuals should not eat yogurt"—fact or fallacy? This happens to be fact. Some individuals, due to the congenital deficiency of an enzyme (galactose-l-phosphate-uridyl transferase), may develop cataracts if they eat foods containing high levels of galactose. Many commercial yogurts—made with milk from which much of the butterfat has been removed, and to which dry milk has been added—happen to be high in galactose. Although the induction of cataracts in such enzyme-deficient individuals by high levels of galactose in the diet had been reported as early as 1935, a stormy controversy broke only as recently as 1970. It was produced by the report of a laboratory study in which twenty-six rats, kept alive on a diet exclusively of yogurt, ultimately developed cataracts.(15)

Two Johns Hopkins University researchers, Curt P. Richter, psychobiologist and James Duke, opthalmologist, reported their findings of a yogurt study that had actually been conducted fifteen years before, but had somehow been sidetracked. The scientists could no longer remember which commercial brands of yogurt they had used. Nevertheless, the researchers were concerned about persons on heavy diets of commercial yogurt. Dr. Richter stated, "I'd rather they drew their own conclusions, but it might be dangerous for people whose diet is made up in large part of yogurt. Over a period, it might have some effect on the eyes."(16) On what did he base this judgment?

Originally, the Richter-Duke experiment was begun as an attempt to learn why Mediterranean people frequently have peritonitis—an inflammation of the stomach lining. They noted that fermented milk, notably yogurt, was a basic part of the diet for many Mediterraneans. Thus the rat experiment was launched.

Rats were found to accept yogurt avidly, taking only small quantities of water with it. An exclusive yogurt diet had no effects on reproduction or activity. After several months, however, all of the animals fed solely on yogurt developed cataracts in both eyes, but without any other sign of toxicity. In young rats, the initial streaks or lines that signal the beginning of cataracts appeared after the animals were on the diet for twenty-eight to forty-two days, whereas in adult rats, this appearance took from sixty-eight to eighty-three days. The rate of maturation of cataracts was less clearly related to age, but in adults, mature cataracts generally developed after four to six months on the all-yogurt diet. The lack of other toxic signs indicated that dietary deficiency of riboflavin or tryptophan would not explain this effect. The clue was found in the unusually high galactose content—from 22 to 24 percent—of the commercial samples of yogurt used. This type of yogurt was made of milk from which the butterfat had been removed, and to which skim milk powder had been added to give an acceptable consistency to an otherwise thin and watery product. The galactose content, which is far less in yogurt made from whole milk, was thus brought into the *known* cataractogenic range. The yogurt-induced cataracts could not be distinguished from those induced experimentally by galactose. This observation confirmed the diagnosis.

After further evaluation, Drs. Richter and Duke concluded that the normal day-to-day ingestion of commercial or homemade yogurt by a majority of the population, could still be considered *without* hazard.(17) However, it is possible that under extreme conditions of use, yogurt could produce cataracts in man. In this connection, it is interesting to note that in parts of India, where de-fatted yogurt forms a large proportion of the diet, the incidence of cataracts is very high. A careful study, however, would be required to establish whether or not defatted yogurt plays any role in the development of these cataracts.

Discussions of various aspects of the galactose-intolerance problem followed the Richter-Duke report. Margarita Nagy, from the American Medical Association's Department of Foods and Nutrition, made some important points, one being that while yogurt is usually only a small fraction of the human diet, in this experiment it figured as the sole food given to the rats. Also, the metabolic system of the rats is different from that of men, in that the rats have an enzymatic deficiency which does not allow them to tolerate lactose and galactose. Most humans have an ample supply of this enzyme—except for some people who have a congenital deficiency and therefore are prone to develop cataracts with the ingestion of foods high in galactose. Side effects, however, usually do not result, due to the early detection of the deficiency. A mono-diet of any food is dangerous. Foods of all the Four Basic Food groups should be eaten, and as part of a *balanced* diet, yogurt is particularly nutritious. The galactose content of commercial yogurt is usually 1.5 to 2 percent. Cataracts beginning to develop in rats on yogurt diets will regress if yogurt is eliminated.(18)

The editors of *Food and Cosmetic Toxicology* made some observations. It is certain that the ingestion of more galactose than can be metabolized, may induce cataracts. Yet, even a large intake of yogurt may not adversely affect the human, since galactose content in yogurt made without skim milk supplementation is comparable to that of normal, whole milk.(19)

Now that the controversy has subsided, it can be seen that the Richter-Duke research has considerable merit. It points up a danger that needs to be recognized: the hazard of modern food processing, which may produce profound changes in staple foods. The modification of milk in commercial yogurt, or the modification of milk by homogenization, or the modification of oil by hydrogenation are a few illustrations of processings that need to be carefully examined in terms of their profound effects on nutrition and health.

◆ ◆ ◆

"Use of fermented milk products can lead to dental decay." This is a fact, in the same sense that any acid food or drink, left in the mouth without proper tooth brushing or mouth rinsing can result in dental decay.

Present evidence demonstrates that dental decay results from the production of acid by bacteria in the plaque, especially bacteria belonging to the *L. acidophilus* type. The acid is produced in the plaque as it adheres to the surface of the tooth. It is not washed away by the saliva, but gradually attacks the dentine. Unchecked, it finally exposes the sensitive pulp. The production of acid is facilitated by the build-up of a complex carbohydrate in the plaque. What the acid-producing bacteria like best in the plaque is the particularly complex carbohydrate called dextran. This can be built up from any sugar, chiefly by *Streptococci* bacteria, but very much more is produced from sucrose (sugar).(20)

Therefore, although *L. acidophilus* and other *Lactobacilli* are highly desirable in the intestine, they are undesirable as residues in the mouth. Proper tooth brushing and mouth rinsing should be practiced after any fermented milk products have been ingested. This is especially important if they are used just before nightly retirement.

Actually, there are many other acids present naturally in foods and beverages, as well as others added by processors. Acids are present in many fruits and fruit juices. A very large number of acids are used to acidulate processed foods. These include such acids as fumaric, malic, succinic, tartaric, glucono-delta lactone, sulfuric and phosphoric. To various degrees, *all* acids make the teeth susceptible to damage.(21)

♦ ♦ ♦

There is at least one occasion when fermented milk products may be *counterindicated* for therapy. Is this a fallacy or a fact? According to Edmé Régnier, M.D. this is a fact. Dr. Régnier believes in the efficacy of therapeutic doses of vitamin C to avoid the common cold, or alleviate its symptoms. He advises that in order to use this therapy, the patient must not have eaten a large amount of yogurt, buttermilk, or other fermented milk products, eight hours prior to the beginning of the treatment. He explains that many people use fermented milk products to avoid constipation. "One way in which they do keep things moving along nicely in the gastrointestinal tract is by forming curds, and these masses of digesting milk-foods stimulate the intestine to contract and push things along in the proper direction."

But the curds have a shape and consistency so that they can absorb other things that have been eaten. Because of this, if yogurt is present in the intestine, any vitamin C taken to control a cold may be detained from reaching the blood stream and prove to be ineffective.

Dr. Régnier advises that yogurt, buttermilk, and other fermented milk drinks be eliminated if you are trying to treat a cold which has developed. If you happen to have eaten a large quantity of these foods at the time of the cold's outbreak, it will be difficult to know exactly how much of an effect this will have in making the vitamin C therapy ineffective. But the longer the period since you ate any of these curd-forming products, the greater the possibility that they will be far enough along in the digestive tract so that they will not interfere too much with the vitamin C absorption, and it will still be able to function effectively.(22)

Notes

1. "Yogurt Challenged as a Medicine," *Consumer Reports*, Oct. 1962, p. 479.
2. Elizabeth Alston, "Yogurt—fad or first-rate food?" *Family Health*, May 1972, p. 22.
3. *Ibid.*, p. 23.
4. *Ibid.*, p. 23.
5. "Cultured Dairy Foods." Table I, Proximate Composition and Mineral Content of Cultured Dairy Foods Compared with Milk. *Dairy Council Digest*, July-Aug., 1972, Vol. 43, No. 4, p. 20.
6. John E. Cooney, "Oh, No, Not Yogurt! Many People Hate It, Yet Many Now Eat It." *The Wall Street Journal*, March 28, 1972, p. 1.
7. Barbara Kraus, "*Calories and Carbohydrates*, New York: Grosset & Dunlap, Inc., 1971, pp. 315-316; *Composition of Foods, USDA*, Agriculture Handbook No. 8, 1963, revised, p. 120.
8. Elizabeth Alston, *Op. cit.*, p. 23.
9. Beatrice L. Gold and Paul Samuel, M.D., "Effect of Yogurt on Serum Cholesterol," *Journal of American Dietetic Association*, Vol. 47, No. 3, Sept. 1965, pp. 192-193; abstracted in *Nutrition Notes*, Oct. 1965, No. 39, p. 5.
10. *The 75th Report on Food Products*, Bulletin 719, Connecticut Agricultural Experiment Station, 1971, p. 60.
11. Elizabeth Alston, *Op. cit.*, p. 22.
12. "Yogurt Challenged as a Medicine," *Op. cit.*
13. *Science News*, May 13, 1967.
14. John E. Cooney, *Op. cit.*
15. C. P. Richter and J. R. Duke, "Cataracts Produced in Rats by Yogurt," *Science*, June 12, 1970, Vol. 168, p. 1372.
16. Richard R. Leger, "Now Yogurt is Struck by a Health Dispute; Tests on Rats Cited," *The Wall Street Journal*, June 15, 1970.

17. C. P. Richter and J. R. Duke, "Yogurt-induced cataracts: comments on their significance to man," *Journal of the American Medical Association,* 1970, Vol. 1878, p. 1878; abstracted in *Food and Cosmetic Toxicology,* Dec. 1971, Vol. 9, No. 6, p. 922.
18. Margarita Nagy, "Questions and Answers," *Journal of the American Medical Association,* Aug. 23, 1971, Vol. 217, No. 8, p. 1113.
19. *Food and Cosmetic Toxicology,* Aug. 1971, Vol. 9, No. 4, p. 603.
20. John Yudkin, M.D., *Sweet and Dangerous.* p. 133. New York: Peter H. Wyden, 1972.
21. "Effect of Acids on Teeth." *Consumer Bulletin,* Feb. 1972, pp. 22-24.
22. Edmé Régnier, M.D., *There Is a Cure for the Common Cold!* pp. 95-96. West Nyack, New York: Parker Publishing Company, 1971.

YOGURT
How to Culture It; How to Use It

"NO ONE made richer, smoother yogurt than my Little Aunt of blessed memory," reminisced Turkish-born Selma. In the afternoon, after her other chores were finished, Little Aunt turned to the ritual of yogurt making. She reserved a special room. She worked alone and undisturbed. It was a rare occasion when she allowed any of the children to watch, and they had to promise to behave themselves. Selma continued, "I can still see the small, tidy room. . . . In one corner was a wooden box filled with straw with its cover nearby and numerous old blankets piled neatly on the shelf."

Rows of earthenware bowls lined the counter. The insides of the bowls were brown, and the outsides were adorned with beautiful green glaze.

Little Aunt would pour milk into a pot reserved exclusively for boiling the yogurt milk. She removed it when the milk came to a boil.

"Rule number one, she used to say, never overboil the milk; the yogurt will have a bad taste," recalled Selma.

The hot milk was poured into the bowls and allowed to cool. That, Selma was told, was rule number two. In a small bowl Little Aunt had some yogurt, kept from the previous day, which would be used to thicken the fresh milk. Impatiently, Selma would ask Little Aunt how she knew when the milk had cooled enough.

"I put my little finger in the milk, and if it does not burn me, I know it is ready." When Little Aunt was satisfied that the critical time had arrived, she added a spoonful of the old yogurt to the fresh milk. She carried the bowls to the box, set them inside, piled straw around them, placed the cover carefully and laid blankets on top.

"Rule number three," announced Little Aunt, while rearranging the blankets and tucking in their edges, "the place has to be warm. Otherwise, the yogurt will not set." On cold wintry days, as an added precaution, Little Aunt instructed that the box should be carried to a place near the coal stove in the living room.

Only Little Aunt knew when the yogurt was ready. She would lift the covers gingerly, look inside, and then her face would break into a satisfied smile. She would announce with pleasure in her voice that the operation was successful.

"Somehow no yogurt tasted as good as the kind my Little Aunt prepared with care and love, and served with pride when we all gathered around the big dining-room table,"(1) recalled Selma nostalgically.

Through the centuries, many people have prepared homemade yogurt using techniques similar to Little Aunt's. The strains of culture may not always have been pure, the heating of the milk was not highly accurate, and absolute control of the incubating temperature was difficult. Yet despite these problems, good yogurt was made.

Ben Bagdikian, of Armenian descent, described the method used by his people: ". . . the proper temperature for the start of the incubation period is ascertained by dipping the elbow into the milk until one feels a sensation of neither hot nor cold. Then the covered bowl is wrapped, first, in the second section of *The New York Times* and over that a baby blanket made of virgin wool given by an aunt in Watertown, Massachusetts 12 years earlier. The coverings are kept in place by a flat granite rock weighing approximately eight pounds. The next morning you have yogurt fit for an Armenian."(2)

Although the basic techniques of the Armenians and the Turks, as well as those of other peoples, remain intact, modern day commercial yogurt had added the advantages of pure strains, temperature control, stainless steel vessels, dairy thermometers, and other paraphernalia that assure quality control.

Commercial yogurt is made by adding additional milk solids in the form of condensed skim or nonfat milk powder to fat-adjusted whole milk. The fortified milk is then pasteurized, cooled, and inoculated with a yogurt culture. Although the specific cultures are not usually included in the label information, yogurt *should* contain *L. acidophilus, L. bulgaricus,* and *Streptococcus thermophilus.* They thrive best when combined, rather than singly.

After inoculation, the yogurt culture is allowed to grow until the proper degree of acidity has been developed. Then it is chilled to firm the body so that it can be handled without separation.

Commercial yogurt *should* be dated, and used promptly. Although it will keep from perishing for a long time under refrigeration, its bacterial value is of a shorter duration. After about a week's time, yogurt becomes what the American Medical Association termed merely "an expensive sour milk."

To date, the Food and Drug Administration has not established any Standard of Identity for yogurt. Therefore, all ingredients used in a particular yogurt must be listed on the label. Although traditional yogurt was made solely with whole milk, most commercial yogurt in the United States is not. Generally it is made of partially skimmed milk, with added nonfat dry milk powder to give a richer flavor and thicker texture, and to add extra protein and calcium. To this basic milk mixture, some dairies add dextrose, a form of sugar for sweetening; and cornstarch, gelatin, guar gum or carrageenan to give the finished product a firmer texture. Carrageenan, one of these thickening agents, is now under close scrutiny by the FDA. Although it is a natural substance, a seaweed, and has been widely used in foods and medicines, it has displayed some adverse effects in its degraded form. It has caused ulcerative colitis in some species of test animals, under certain conditions. The lesions appear similar to those found in human ulcerative colitis.(3)

In addition to the thickening agents, many of the commercial yogurts have flavorings. Some fruit-flavored yogurts have preserves at the bottom of the container; Swiss-style yogurts have fruit and sugar blended throughout the products.

Many people, especially in areas where it is difficult to find good-quality yogurt, have come to prefer to make their own yogurt at home. Homemade yogurt can be made as mild or tart as one wishes. Homemade yogurt can be made and eaten when fresh—with no guesswork regarding its age. Homemade yogurt can be flavored without adding sugar or sugar-laden ingredients. Homemade yogurt can be made easily, and save money. "You'll like saving 70 percent of the cost of commercial yogurt"—according to the statement of a large manufacturer of an apparatus for making homemade yogurt.

The large number of yogurt-making devices now being sold for home use confirms the trend toward homemade yogurt. Some of these devices rely on electricity, while a few do not. When directions are followed explicitly, these devices usually make foolproof yogurt. However, commercial yogurt makers are *not* essential.

If you plan to make yogurt without a commercial device, there are numerous types of commonly available pieces of equipment that many people have used successfully. Choose what you prefer, and approach the activity with a sense of experimental investigation. You may find that you prefer one method to another. Explore various possibilities until you settle upon one that works well for *you*.

Milk: Any type of milk will make yogurt. But do not use condensed milk. It can be raw, pasteurized, homogenized or low-fat; from cow, goat, mare, sheep, buffalo, or soy. The milk should be fresh. The older the milk, the longer it will take to incubate. Or, you may need to use more starter. If you have access to clean raw milk, morning milk is said to make the best flavored yogurt. Raw milk needs to be pasteurized for making yogurt, in order to kill other bacteria which would interfere with the *Lactobacilli*.

Starter: Use a good-quality, fresh commercial starter, or save some from your own batch. Always use plain, unflavored yogurt as a starter. In the Mediterranean area, yogurt is sometimes dried and powdered, and kept for a period of time. Such a mother culture is known as the "Podkwassa" or "Maya."(4)

Temperature control: Although a candy or dairy thermometer may take the guesswork out of yogurt making, neither are indispensable. Nor is it necessary to use the time-honored practice of sticking a thumb or an elbow into the heated milk. Simply sprinkle a few drops of milk on the wrist—as you would test milk in a baby's bottle. If you use an electric stove to heat the milk, note the amount of time required at a particular dial setting to have the milk reach the proper temperature. Once this is known, use the same time and dial setting for subsequent batches. If you use a gas, coal, kerosene or woodburning stove, this kind of information may be trickier to determine. However, once you gauge the time and heat required, you can make this step routine by setting a kitchen timer or alarm clock.

Cooking vessel: Use any inert material, such as stainless steel (unpocked), enamelware (unchipped), glass or Corningware.

Yogurt containers: These, like the cooking vessels, should be made of an inert material. They should not be cracked, crazed, or have poor glazes. They may be widemouth glass canning or freezing jars, ovenproof glass or crockery custard cups, earthenware bowls, stone crocks, or casseroles. If the containers do not have lids, improvise. Some plastic lids from cans of peanuts or coffee, for example, snap tightly over some custard cups. Lids can be cut from circles of brown paper, parchment paper or aluminum foil, and held in place with rubber bands or string.

Incubators: Many yogurt makers have created ingenious and inexpensive incubators from a variety of materials. The following suggest the richness of possibilities.

An inexpensive polystyrene ice bucket or picnic hamper. If you make a small quantity of yogurt in small jars, the former would be satisfactory; if you make a larger amount of yogurt, the latter is preferable.

An electric skillet, or an electric heating pad turned to the lowest heat. Some people wrap the jars in towels, place them in an insulated bag, and insert the electric heating pad at the bottom of the bag.

An electric stove oven heated to 120° F., and then turned off. Allow the oven to cool to 90° F., and, while the incubating yogurt is in the oven, keep the oven temperature between 90 and 105° F.

An iron pot, containing warm water. Lower the jars of incubating yogurt into the pot, and cover with a blanket; check the temperature of the water from time to time, and reheat if necessary.

Place the yogurt in a double boiler over a gentle source of heat: such as the pilot light on the top of a gas stove, a steam-heat radiator, a hot-air floor register, a banked-woodstove fire or the warm area at the back top of a refrigerator.

A widemouth thermos.

An old-fashioned "featherbed comforter" or a warm blanket. Use as a wrap for the incubating yogurt.

An old-fashioned "fireless cooker" or "haybox." T.W. Fowle has kindly supplied the following directions: "For a family-sized batch, make a corrugated paperboard box seven inches high by seven to eight inches square, enough space for four tall "Skippy" jars. Quart jars require a box at least eight inches square. Tuck thick folded huck towels over and around the jars, then cover completely with a heavy blanket. I put this small box into a larger one, twelve by twelve inches, and fifteen inches high. I insulate between the two boxes, at the bottom, top, and sides, by using clean cotton, cotton rags, shredded paper or cotton batting. Bend the top flaps of the smaller box downward, so that they can be tucked into the larger box, and provide additional insulation on the sides. To cover the inside box completely, place a pillow on top. Close the top of the outside box, and hold it in place with heavy cord. A piece of plywood placed under the cord will keep the top flat and closed, and protect the box from being cut into by the string."

Instead of using cardboard boxes, you can use wooden ones. For insulation, try vermiculite, sawdust, wood shavings, or feathers. Or try the excelsior or the polystyrene chips frequently used when fragile objects are shipped. All of these materials are good insulators. However, be careful to keep them out of the incubating yogurt.

Some do's and do not's for yogurt making:

OBTAIN a good-quality starter if you are making yogurt for the first time. Special starters can be purchased. Or, use a good-quality, fresh, plain commercial yogurt. If, as your yogurt making progresses, you notice that the product begins to falter, renew the starter, or give it a "boost" with additional starter.

INCUBATION TIME will vary, depending on several factors, including the freshness of the milk, the temperature of the culturing, and the strength of the starter. In general, two to three hours of culturing, at a temperature between 105° and 112° F. seem to be most favored by yogurt experts. The quicker the milk cultures, the milder the taste, and the greater the number of beneficial bacteria. The longer the milk cultures, the more tart the taste and the lactic-acid content.

In order not to overincubate, it is preferable to make yogurt during the daytime rather than have it culture overnight. Overincubation toughens the curd. This is undesirable, and makes difficult the straining of the yogurt which is done before putting it into a baby's bottle.

Do not use too much starter, or you will get lumpy yogurt. A tablespoon of starter for each quart of milk is enough to culture it. Make certain to mix the starter thoroughly with the milk, or you may end up with a container of milk with a dollop of starter at the bottom.

If the milk becomes too cool during the incubation period the growth of bacteria will be retarded. If this has happened, adjust the temperature, incubate longer, and if necessary, add a little more starter. If the milk has become too hot (above 115° F.) the yogurt-making bacteria will be killed off. But the too-hot milk need not be discarded. Instead, turn it into cottage cheese. (For directions, see pages 91-92.)

Exercise care that all containers and utensils used in making yogurt are scrupulously clean. It is a good practice to scour them with boiling water.

Before turning the warmed milk into containers, rinse them in hot water. This practice will help to keep the milk at a good temperature.

Do not transfer the culturing milk to another container. Allow it to remain, undisturbed, while it is incubating. In checking to see if the milk has cultured, tip the container gently, and slightly. The watery separation of yogurt is merely whey, which is nutritious. Never attempt to stir the whey back into the yogurt. This is a sign that you are a yogurt novice, and you will get glares from the yogurt aficionado. To minimize the separation, use a wooden spoon, or some other light-weight implement. Rest it lightly on the surface of the yogurt, and scoop the yogurt from the top.

IF YOU WISH to thicken yogurt with dry milk powder, blend both together before heating the milk. Use anywhere from two tablespoons to four tablespoons of dry milk powder for each quart of milk.

REMEMBER THAT the cooling cultured yogurt will continue to thicken somewhat after it is refrigerated.

IF YOU PLAN to flavor the yogurt, add the flavor to the finished product.

NEVER USE flavored yogurt for culturing the subsequent batch.

STORE YOGURT in the refrigerator between 35° and 45° F. Remember, the fresher the yogurt, the higher its bactericidal value.

How to use yogurt:

Yogurt is best used fresh, before it is more than a week old. It should not be subjected to further heating. The bactericidal value is lost when yogurt is heated above 115° F. It is best to plan to use yogurt, whenever possible, in its natural state, or with fruit, in salad dressings, dips, cold soups and sherbets. It can be added to the top of baked potatoes in lieu of sour cream, or spooned over other cooked vegetables. It can be used as a garnish on top of hot soups, casseroles and baked apples.

If you do not like plain yogurt, try some of the following flavorings: cinnamon, nutmeg, unsulfured molasses, honey, carob, pure vanilla extract or pure almond extract. Or, top yogurt with ground nuts, sesame seeds or wheat germ. Yogurt combines very well with all fruits, berries and melons. Soaked, dried fruits such as prunes, dates, figs, apricots, peaches, apples or raisins are particularly good with yogurt, if you object to its normal tartness.

Yogurt salad dressing: Blend together in an electric blender a cup of homemade yogurt, ⅔ cup safflower oil, ⅓ cup apple cider vinegar, an onion quartered, ½ garlic clove, ¼ cup celery leaves and ¼ cup parsley leaves. Yields 2½ cups of salad dressing.

Yogurt dip: Blend together in an electric blender a cup of homemade yogurt, two avocados, a garlic clove, and three tablespoons of minced chives. Yields a pint of dip. Serve on thin rounds of raw white turnips.

Yogurt-tomato soup: Blend together in an electric blender a quart of tomato juice, a pint of homemade yogurt, and a sprig of fresh tarragon. Chill. Serves six.

Yogurt-cucumber soup: Blend together in an electric blender a pint of homemade yogurt, three peeled and coarsely chopped cucumbers, the juice of one lemon, a sprig of fresh dill, a garlic clove, three tablespoons of safflower oil and a tablespoon of fresh mint. If you wish to make a smooth-textured soup, remove the seeds from the cucumbers before putting them into the blender. Chill. Serves six.

Yogurt sherbet: Blend together in an electric blender 1½ cups of soaked and drained dried apricots or peaches with 1½ cups of homemade yogurt. Turn this purée into six sherbet glasses. Chill. Serve garnished with ground sesame seeds. Serves six.

These recipes are among many which are readily accepted and enjoyed by people who *think* that they loathe yogurt.

In some countries butter and ghee are prepared from yogurt, by heating it until the water content is lowered from about 20 percent to less than 1 percent. A study was made of the chemical and biologic properties of butter and ghee made from yogurt, contrasted to similar products prepared from sweet milk. Butter and ghee prepared from yogurt had a higher total acidity and biacetyl value, as well as a more intense and sharper flavor than the fat samples prepared from whole milk. There was also a difference in flavor between butter and ghee made from yogurt. This was associated with a lower biacetyl value in ghee, attributed to the longer heat treatment required to reduce the water content.

One reason for the popularity of ghee is that it keeps well. After two weeks' storage at room temperature, butter samples had to be discarded because of mold growth, whereas the ghee showed no signs of mold. Not only the higher water content of butter, but also the presence of other substances such as protein curd are thought to make it more favorable for mold growth than ghee.(6)

On the occasions when you wish to use yogurt in cooking, remember the following:

Yogurt, substituted for sour cream is less rich, but more tart.

Yogurt, substituted for buttermilk, in biscuits or pancakes, can be satisfactory. For each cup of buttermilk called for in the recipe, use one and one-half cups of yogurt. If baking powder is called for in the recipe, substitute one-half teaspoon of arrowroot flour to each cup of yogurt.

Cook yogurt at low temperatures, for a brief duration, and stir constantly. This will help to prevent separation. Cooking in a double boiler is advisable.

Yogurt may lose its flavor when cooked, but it still has the ability to intensify and blend the flavors of other foods. Yogurt is often used to blend herb flavors into sauces and gravies, or to marinate meat.

Notes

1. Selma Ekrem, "Little Aunt's Yogurt," *Christian Science Monitor*, May 4, 1970.
2. Ben Bagdikian, letter to the editor, *Science*, Nov. 6, 1970, quoted by Natalie Ganley, in "Yogurt Making," *Washington Post*, July 20, 1972.
3. Beatrice Trum Hunter, "Carrageenan, an Additive Widely Used in Food Manufacture–A Cause of Human Ulcerative Colitis?" *Consumer Bulletin*, May 1972, pp. 16-17; 34.
4. "How to Make Yoghurt," USDA Agricultural Research Service, Dairy Products Laboratory, *Publication CA-E-15*, 1967.
5. R.I. Tannous and A. Merat, "Chemical, Physical and Biologic Properties of Butter and Ghee," *Journal of American Dietetic Association*, 1969, Vol. 55, pp. 267-272; abstracted in *Dairy Council Digest*, Jan.-Feb. 1970, Vol. 41, No. 1, p. 6.
6. *Ibid.*

KEFIR
How to Culture It; How to Use It

KEFIR is a rather well-known fermented milk drink long enjoyed throughout the Balkan countries. Moslems in the Caucasus thought that the ferment of kefir, or "Grains of the Prophet Mohammed" would lose its strength if people of other religions would use it. For this reason, the preparation of kefir remained a secret for a long time. It is thought that the word "kefir" came from keif, a Turkish word meaning "good feeling," for the sense of well-being experienced after drinking this fermented milk product.

Outside of the Caucasus, kefir was scarcely known. Marco Polo mentioned kefir in his account of his Eastern travels. After that, for nearly five centuries, kefir was forgotten in the West. Renewed interest in kefir was displayed in the early nineteenth century when it was used therapeutically for tuberculosis at lung sanatoriums.(1) Today, many of the qualities in kefir have become recognized as being similar to those in yogurt; however, kefir has some unique properties.

Kefir "grains," the fermenting agent, are convoluted gelatinous particles obtained from fermented milk of a type commonly prepared in the countries of southwestern Asia. In appearance, they have been described as looking like cauliflower pieces, or coral-like. The kefir grains vary in size from the size of a grain of wheat to that of a hazelnut. They contain three kinds of lactic acid bacteria, and lactose-fermenting yeasts.(2)

What are the unique properties of kefir? Kefir has a very low curd tension. This means that the curd breaks up very easily into extremely small particles. (The curd of yogurt holds together or breaks into lumps.) The small particle size of the kefir curd facilitates its digestion by presenting a large surface for the digestion agents to work on. This ease of digestion has led to many researchers recommending kefir as a food particularly beneficial to infants, convalescents, elderly persons, or persons with insufficient digestive activity.(3)

It has been found that kefir stimulates the flow of saliva, probably due to its acid content and its slight amount of carbonation. Kefir increases the flow of digestive juices in the gastrointestinal tract, and stimulates peristalsis. Because of this latter virtue, kefir has been recommended as a post-operative food, since many abdominal operations cause the temporary cessation of peristalsis, accompanied by gas pains.(4) A noted Danish dairy bacteriologist, Dr. Orla-Jenson, expressed the opinion that kefir has a *higher* nutritive value than yogurt due to the abundance of yeast cells digested and the beneficial effect on the intestinal flora.(5)

In addition to the above virtues, kefir is different from yogurt in at least two other respects. Kefir is of a thin consistency, and generally is drunk, like buttermilk; yogurt is usually of a thicker consistency, and generally is eaten. In preparing milk for yogurt, it must be heated in order to destroy unfavorable microorganisms that would otherwise interfere with the fermentation. In preparing milk for kefir, the milk does *not* need to be heated. Hence, if one has access to Certified Raw Milk or a source of clean fresh raw milk, he has an ideal medium for culturing kefir.

In other respects, kefir has as many similar nutritional qualities and physiological effects as other cultured milk products. Kefir is laxative, and in Germany and in the Soviet Union it is used extensively with cases of chronic constipation. Kefir, like yogurt, is used for a wide variety of intestinal disorders.(6) Kefir, like yogurt, also displays bactericidal powers against some virulent pathogenic organisms, and is also useful for reestablishing the "friendly flora" in the intestine after antibiotics have been administered.

Kefir grains are available in several forms. They are shipped commercially through the mail, suspended in milk.* They are also sold in freeze-dried form.** Although attempts have been made to process them in tablet, cake, and wafer forms, to date, such processed products appear to lose much of their value since the heat above 120° F. used in the processing, destroys lactic acid bacteria.

Homemade kefir: Obtain a supply of fresh kefir grains, suspended in milk. Culturing them is very simple. Add the kefir grains to whole or skim milk, stir, cover, and allow the mixture to remain at room temperature (65° to 76° F.) for two to three days. At the end of this time, the milk will have thickened. Pour it through a sieve. What drains through is the drinkable kefir. What remain in the the sieve are the active kefir

*Active kefir grains can be ordered from R-A-J Biological Laboratory, 35 Park Avenue, Blue Point, New York, 11715.

**Freeze-dried kefir grains can be ordered from Continental Culture Specialists, 1354 East Colorado Street, Glendale, California, 91205.

grains, which can be used, over and over again, to culture future batches. Before transferring the grains to a new supply of milk, rinse them thoroughly in a sieve, using cool, running water. Drain off the excess water, and add the washed grains to the new batch of fresh milk you wish to culture. The proportion of grains to milk should be about one teacup of kefir grains to one quart of milk. As you continue to culture, the number of kefir grains will thrive and multiply. This increase takes place most rapidly when the kefir grains are cultured in skim or low-fat milk (less than 1 percent butterfat held at room temperature (68° to 70° F.).

If you wish to slow down the multiplication of kefir grains, hold the culturing milk at a lower temperature (55° to 60° F.). If you do this, you will only have to make batches of kefir weekly instead of every second or third day. Your choice may be determined by the amounts of kefir you wish to consume.

When your grains multiply, and you find you have more than a teacupful for each quart of milk, you will have to come to a decision. You can allow the abundance of grains to remain, which will result in a thicker product. In time, the extra grains will have to be removed. You can give them to a friend, neighbor or relative who is anxious to begin culturing kefir. (Such extra grains are excellent fund-raising items for worthy causes.) Or, you may decide to preserve the grains for future use (see below). Or, you can culture more kefir than you can use as a beverage, and turn the surplus into kefir cheese (see below).

In culturing kefir, you can control the flavor as well as the consistency. If you enjoy it tart, culture for a longer period of time; if you like it mild, culture more briefly. If you enjoy thick kefir, allow a large quantity of grains to remain in the culture; if you like thin kefir, remove the excessive grains as they multiply.

Preserving kefir grains for future use: If you plan to be on vacation and have no "kefir sitter" to nurture the grains in your absence, preserve them for future use. Store them wet, dry or frozen.

To store them wet, wash the grains thoroughly in a sieve with a stream of cold, running water. Place them in a clean jar, and cover them with cold *unchlorinated* water. Store in a refrigerator at 40° F. They will keep their viability for about a week, but lose potency if held much longer.(7)

To dry the kefir grains, rinse and drain as above. Place them on two layers of clean cheesecloth. Air-dry them at room temperature for thirty-six to forty-eight hours. The room should have good air circulation, or use a fan. Place the dry grains in a paper envelope or wrap them in aluminum foil. Keep them in a cool dry place. Such dried grains are usually still active after twelve to eighteen months of proper storage.(8)

To freeze kefir grains, rinse and drain as above. Place them in a small non-actinic bottle (light-excluding). Seal the bottle tightly and place it in a similar but larger bottle. Seal the edges of the lid of the larger bottle with masking tape to prevent any air from entering. Freeze, and thaw out to begin culturing again at a future date.

Reactivating stored kefir grains: You may find that kefir grains stored wet may be slightly slower in their culturing action. If so, leave them in the fresh milk somewhat longer than usual, or culture them at a slightly higher temperature than usual. In time, they should be restored to their full activity.

To reactivate the dried or frozen grains, first soak the grains overnight, or approximately twelve hours in *unchlorinated* water to cover, at room temperature. Drain, and transfer the grains to only *one cup* of milk. Allow this to stand, approximately twenty-four hours, at room temperature, or until the milk thickens. During this period, stir the mixture occasionally. Drain the grains from the milk, rinse, and transfer them to a fresh cup of milk. By then, the grains should be fairly active. They can be rinsed, drained, and transferred to *two cups* of milk. Continue, gradually increasing the number of cups of milk used, until ultimately the usual one quart of milk can be cultured with the grains. Continue to culture under the conditions that you have found produce the flavor and thickness which suit your taste.

Homemade kefir buttermilk: Kefir grains can be used to prepare kefir buttermilk as well as to prepare the usual type of kefir fermented milk. Kefir buttermilk closely resembles churned or real buttermilk in flavor and consistency and has proved to be a popular item when it has been sold along with cultured buttermilk. Use either raw or pasteurized partially-skimmed milk (two to three percent butterfat content), or pasteurized sweet buttermilk, which is sometimes available from creameries where butter is made from sweet cream. Of course, it would also be available on a farm where butter is churned from sweet cream.

If raw milk is used, heat it to 165° F. Cool it to between 68° and 72° F. It is then ready for making the kefir buttermilk. Or, it may be kept in the refrigerator for later use. If sweet buttermilk is used, pasteurize it by heating at 180° F. for twenty minutes to expel all the air.

To prepare a small batch of kefir buttermilk, put the kefir grains in a cheesecloth bag. A half pound of grains will be enough to coagulate two gallons of milk in thirty-six to forty-eight hours at 68° to 72° F.

Place the milk in a porcelain, glass-lined or granite-iron container. Suspend the bag of grains near the surface of the milk. Incubate the milk at 68° to 72° F. until it is well set up, or until it shows an acidity of about 1 percent lactic acid. Then lift out the bag of grains, scrape off the adhering cream, and wash the bag thoroughly in cold, running water to remove all the sliminess from the outside of the bag, and the fermented milk from within. After squeezing the bag to remove the excess water, place it in a new supply of milk, and continue as before.

After the grains have been removed, stir the fermented milk slightly to incorporate all of the surface cream. Then cool it to 60° F. or lower. Then, using an eggbeater or other slow-speed type device, whip the mixture to break the curd. After the curd is broken, store the milk in bottles or in its original container at a low temperature until it is used.

Homemade kefir cheese: Pour cultured kefir (without the grains) into a pot, and heat very gently until the entire surface is covered with foam, just below the boiling point. Turn off the heat, and allow the heat to permeate the entire mixture for a few seconds. Then pour into another container for cooling. Cover, and allow to rest overnight. In the morning, strain through a cheesecloth bag. Kefir cheese is thick, and has large curds. To modify its bland flavor, a little grated roquefort or blue cheese, or other strong-flavored cheese, can be blended with it.

Homemade kefir from freeze-dried grains: Sterilize all equipment. Heat a quart of fresh, clean milk to a near boil. Remove the pot from the heat, and cool the milk to 110° F. Stir in a teaspoonful of freeze-dried kefir grains, and blend thoroughly. Pour this mixture to fill a sterilized, warm container, cover, and place in a warm place, such as an oven at 100° F. Allow the container to remain undisturbed for twelve hours, while the kefir is incubating. Test the kefir with a toothpick to learn whether it has solidified. The toothpick, when inserted, should stand upright without support. If the test fails, incubate the kefir one to three hours longer. When the incubation is complete, refrigerate the kefir.

Kefir summer cooler: Blend together in an electric blender a quart of homemade kefir, the juice of two oranges and two tablespoons of honey. Pour into glasses, chill, and top with a light dusting of nutmeg. Serves four to six.

Kefir shake: Blend together in an electric blender a quart of homemade kefir, a half cup of wheat germ, and two tablespoons of honey. Pour into glasses. Serves four to six.

Kefir summer soup: Blend together in an electric blender a quart of homemade kefir, one cup of raw, diced cucumber, one cup of watercress, and a sprig of fresh parsley. Chill and serve. Serves six.

Kefir sherbet: Blend together in an electric blender 1½ cups of homemade kefir, ¾ cup unsweetened, puréed fruit, 4 tablespoons honey, and 1 teaspoon of pure vanilla extract. Pour into sherbet glasses, and refrigerate. Serve, garnished with unsweetened coconut shreds. Serves four to six.

Quickie kefir dessert: Fold one cup of kefir into one cup of applesauce, and add ¼ teaspoon of pure almond extract. Pour into sherbet glasses, and refrigerate. Serve, garnished with ground walnuts. Serves four to six.

Baked apples with kefir: Wash and core raw apples. Arrange them in a baking dish with a small amount of water in the bottom. Fill the core cavities with raisins and almonds. Bake in a moderate oven for a half hour, and serve hot, with cold kefir drizzled over the apples.

Kefir fruit salad dressing: Blend together in an electric blender one cup of homemade kefir, one tablespoon of honey, two tablespoons of unsweetened orange juice and two tablespoons of unsweetened pineapple juice. This dressing is good over fruit salad, and also over cole slaw. Makes 1½ cups.

Kefir dip: Mince a clove of garlic, and mix it with a teaspoon of safflower oil. Add a half cup finely chopped walnuts, a teaspoon of apple cider vinegar, and one cup of kefir. Mix well and chill. Serve with thin slices of raw white turnips.

Notes

1. Clifford W. Hesseltine, "A Millennium of Fungi, Food, and Fermentation," *Mycologia,* Mar.-Apr. 1965, Vol. 57, No. 2, pp. 149-197; José M. Rosell, M.D., "Yogurt and Kefir in their Relation to Health and Therapeutics," Historical Notes, *Canadian Medical Association Journal,* 1932, Vol. 26, pp. 341-345; Harry Seneca, M.D., et al, "Bactericidal Properties of Yogurt," *American Practitioner and Digest of Treatment,* Dec. 1950, Vol. 1, No. 12, pp. 1252-1259.
2. José M. Rosell, M.D., *Op. cit.*
3. L. A. Burkey, *Directions for Making Kefir Fermented Milks.* BDIM–Inf. 58, USDA, ARA, Bureau of Dairy Industry, Nov. 1947.
4. *Ibid.*
5. Orla-Jensen, *Dairy Bacteriology,* 1935; abstracted by L. A. Burkey, *Op. cit.*
6. José M. Rosell, M.D., *Op. cit.*
7. Frank Kosikowski, *Care and Drying of Kefir Grains.* Dept. of Dairy and Food Science, Cornell University, undated.
8. L. A. Burkey, *Op. cit.*

CHEESES
Fermented Milk Products

CHEESE has been described as "milk's leap to immortality." Cheese bears the same relationship to milk that wine bears to grapes. And, like wine, cheese has countless varieties. Because no two grasses, waters or climates are identical, no two cheeses, however similar, can be identical if they are made in two different places. Skilled cheesemakers have attempted to make the same cheese in new localities, but find that such duplication is impossible. However, some of the attempted duplications turn out to be superior to the original. At times an entirely new and great cheese has been produced.

Various countries claim the discovery of cheese as their own. There are many legends, told with numerous local versions, but all have certain elements in common. Someone sets out on a journey, taking with him some milk in a leather pouch made from the stomach of a calf. After riding for a time, the traveler discovers that the milk has turned into a palatable sour curd. Rennin, an enzyme from the lining of a calf's stomach, converts milk into curds and whey. It is used almost universally in cheesemaking.

Cheese dates back to the earliest domestication of animals, estimated at least to 9000 B.C. Cheese has been made whenever animals have produced more milk than people can use at the moment in fluid form. Cheese was well-known to the Sumerians by 4000 B.C. and archaeologists have found references to cheese on their cuneiform tablets. Both Egyptian and Chaldean artifacts refer to cheese. It was a staple of biblical times, along with honey and wine. The ancient Greeks had a deity, Aristaeus, who was considered the giver of cheese.

The wicker baskets used by the Greeks for draining whey from the cheese were known as "formos." The word became "forma" in Latin, from which came the Italian word for cheese, "formaggio," and the French word, "fromage." From the Latin "caseus" for cheese came the German "kaese," the Dutch "kaas," the Irish "cais," the Welsh "caws," the Portuguese "queijo" and the Spanish "queso." The Anglo-Saxon "cese" or "cyse" later became "cheese."(1)

Cheese retains much of the food value of fresh sweet (but perishable) milk. Cheese production represented a way to preserve milk for later use. (For the nutrients in various types of cheeses, see Tables I and II, pages 36-39).

Since hard cheese travels well, it was frequently taken to distant places, especially in the knapsacks of soldiers, from the time of Caesar's legions to the armies of Genghis Khan. Marco Polo brought back from his travels a report of what may well have been the earliest method of preserving cheese: "They make provision also of milk thickened and dried to the state of a paste, which is prepared in the following manner.

They boil the milk, and, skimming off the rich or creamy part as it rises to the top, put it into a separate vessel as butter; . . . The "butter" is then exposed to the sun until it dries. Upon going in service, they carry with them about ten pounds for each man, and, of this, half a pound is put, every morning, into a leathern bottle, with as much water as is thought necessary. By their motion in riding, the contents are violently shaken, and a thin porridge is produced, upon which they make their dinner."(2)

Cheeses are produced from the milk of a great variety of animals including cow, sheep, goat, buffalo, camel, ass, mare, llama, reindeer, yak and zebu. To make cheese, the milk must be caused to separate and form curds and whey. This separation has actually begun when milk begins to sour. It can be hastened by means of a bacterial "starter" (an acid substance such as buttermilk or rennet) which curdles the milk.

What kind of cheese will be produced depends on a number of things. The milk may or may not be heated before or during the time when the starter is added. The milk may or may not be skimmed; it may or may not have extra cream added. The cheese may be made from the milk of only one milking, or it may be made from a mixture of evening and morning milks. A few degrees of difference in temperature will produce distinctly different cheeses, as will the way in which the curd is cut. The size and shape of the cutting instrument, and even the motion used in cutting the curd, will result in differences in the finished cheeses.

After the curds are broken up, they may or may not be drained thoroughly. They may or may not be salted. They may or may not be pressed. They may or may not be inoculated, and their surfaces may be treated differently. Some may be ripened in warm moist places, while others may be stored in cool dry ones. Some may rest on mats or in baskets, while others may be left in cheese hoops or under weights for additional pressing. Some cheese may ripen in a few days, while others may require a number of months. All these variables will produce different kinds of cheeses.

◆ ◆ ◆

Homemade soft cheeses: From ancient times, people have been making soft cheeses. Clabbered milk is perhaps the simplest version. Fresh milk is poured into a shallow utensil and allowed to sour at room temperature. Or, the milk can be gently heated or exposed to the sun. By the time it sours, some of the whey has evaporated; the remainder is a semi-dry custardy-like curd.

The famous Devonshire cream is cream that is allowed to rise on top of the milk, and is then set by heating and cooling. It is skimmed off from the underlying skim milk.

Pot cheese is made from heated sour milk. A starter is not needed if the milk is sufficiently sour. Scoop out the curds and drain them in a cheesecloth until the product is quite dry. Or, rennet* can be used as a starter. For every quart of milk heated, add one teaspoonful of rennet. When the curd is formed, strain the mixture through a clean cheesecloth to separate the curds and whey. Tie the ends together and allow

---

*Rennet can be purchased under the name "Junket" which is a product of Salada Foods, Inc. 399 Washington St., Woburn, Massachusetts 01801. It is also sold by C. Hansen's Laboratories, 9015 West Maple Street, Milwaukee, Wisconsin 53214. The Junket tablets are smaller than the Hansen ones, and therefore more suitable for homemade cheese production. A dozen Junket rennet tablets is the equivalent of one Hansen cheese rennet tablet. One Junket tablet is used for each quart of milk.

the whey to drain out. When the curds are nearly solid, either season or moisten the pot cheese with cream.

Cottage cheese can be made with rennet. Dissolve ¼ tablet of rennet in ½ cup of cold water. Combine one gallon of skim milk with ¼ cup of buttermilk or acidophilus milk and heat the mixture to 70° F. Add the rennet solution and stir well. Cover the mixture with a towel and allow it to stand at room temperature from twelve to eighteen hours, or until a smooth, firm curd forms. Cut the curd into ½-inch pieces, using a long knife. Heat the curds slowly over a pan of hot water (as a double-boiler arrangement) until the temperature reaches lukewarm, 110° F. Hold the curds at this temperature for twenty to thirty minutes, stirring at five-minute intervals in order to heat through the curds uniformly. When the curds are sufficiently firm, pour the mixture into a colander lined with clean cheesecloth and allow the whey to drain off. Shift the curds around occasionally by gently lifting the corners of the cloth. After the whey has finished draining, draw the ends of the cloth together, and immerse the sack in cold water. Work with a spoon to free the curds of any remaining whey. Add sea salt and cream (both optional), mix thoroughly and chill.

Cottage cheese can be made by using yogurt or cultured buttermilk as a starter. Sour each quart of lukewarm milk with two tablespoons of yogurt or buttermilk. Mix thoroughly and keep in a warm place until it solidifies in twelve to twenty-four hours. Place the container in a vessel of warm water (double-boiler arrangement) that is approximately 115° F. for one or two hours, until the milk curdles. Pour the mixture into a colander lined with clean cheesecloth, and continue as above.

Cottage cheese can be made by using lemon juice to

help form the curd. Add the strained juice of two lemons to each quart of milk, and allow the mixture to remain at room temperature until it clabbers. Heat the milk slowly in the top of a double boiler, set over warm water, until the whey begins to rise to the top. Allow the milk to rest for ten minutes away from the heat. Then pour the mixture into a colander lined with clean cheesecloth, and continue as above.

Homemade cream cheese can be made by allowing fresh cream to sour at room temperature. This will take about two days. Pour the soured cream into a clean cheesecloth bag and allow the whey to drain. Remove the solid cheese from the bag. Chill, and then form into flat cakes. Serve as a dessert with fresh fruit.

A homemade Neufchatel-type cheese has been suggested by the New York Agricultural Experiment Station at Geneva, New York. The procedure is as follows: Dissolve ¼ of a rennet tablet in ¼ cup of cold water. Heat one quart of whole cow's or goat's milk to 72° F. Add one teaspoonful of buttermilk or other cultured milk as a starter. Add one tablespoonful of the rennet tablet solution to the milk mixture and discard the remainder of the rennet tablet solution. Mix thoroughly. Allow the mixture to stand for eighteen hours at room temperature, until a firm curd has formed. Then pour the mixture into a cheesecloth bag, and allow it to drain in a cool place. The whey will drain off in about twelve to twenty-four hours. When the curd is firm, use one to two teaspoons of sea salt to season each pound of cheese. For variation, add finely ground pimentos or peppers to the finished cheese.

A homemade ricotta cheese can be made using the whey, which is a byproduct in making a homemade grating-type Italian cheese. (See pages 99-100.) Before putting the grating-cheese curd, which is still in the hoop, back into the whey, heat the whey until a coat of cream rises to the top. Then add 2½ gallons of whole milk, stir, and slowly heat the mixture until it reaches the point *just below* boiling, but *not boiling.* Remove the vessel from the heat source. When the curd rises to the top, and tends to draw away from the edges of the vessel, add a scant half cup of full-strength apple cider vinegar. Stir well. The curd will rise to the top. Skim it off, salt to taste and allow to drain for about eight hours. It is then ready to eat, and should be used while very fresh.

The classic French coeur à la crème is a rich dessert made from heavy cream, cream cheese and cottage cheese. It can be made successfully from lower-calorie ingredients and still be delicious. Blend two cups of homemade cream cheese with one cup of homemade yogurt. Pack the mixture into the traditional heart-shaped coeur à la crème mold. Or, lacking the mold, use a cheesecloth-lined colander. Allow the whey to drain out. Chill at least two hours. Unmold carefully and garnish with a pint of strawberries or other fresh berries or fruits in season. This recipe serves six.

Homemade soft cheeses have been popular in many countries. Among many: in Austria, it has been "rahm kaese"; in Finland, "glumse"; in Holland, "may" cheese; in Latin American countries "queso blanco"; and in India "surti" and "surtal."

In attempting to make homemade soft cheeses, it is necessary to read the labels on all commercial dairy products that may be used as basic ingredients. Some of the additives now employed may interfere with the fermentation process. For example, whipping cream may contain sodium alginate and monoglycerides as emulsifiers and stabilizers. Mixtures of cream and milk ("half-and-half," and "coffee cereal special") may also contain these additives. Low-fat milk may contain tapioca, vegetable gums and carrageenan (see page

62) to give the milk greater body and "mouth feel." Some of the processings now employed may also interfere with the fermentation process. For example, sweet fluid milk and cream may be "flash pasteurized" or "ultra-pasteurized."(5) These techniques extend shelf-life for the products, but may interfere with the souring. When such treated products "turn" they are not apt to sour pleasantly but rather to rot, with a disagreeable bitter odor and taste. Fermentation has not taken place; putrefaction has.

For cheesemaking, the milk quality is important. Good cheese cannot be produced from poor milk. Milk which is already noticeably sour in taste should not be used for cheesemaking. Overripe milk will result in cheese with an unpleasant sour flavor.

It is important that cleanliness be observed in each step of cheesemaking: handling the milk, cleaning and scalding of utensils, and protecting the ripening cheese from insects.

Cheese made from milk which has been handled improperly is apt to be gaseous, with an acid, bitter, fruity or other undesirable flavor. If the curd has not been allowed to firm properly, the cheese may be pasty in consistency, as well as sour in flavor. If this happens, in subsequent cheesemaking, cut the curd into smaller pieces and warm the mixture longer.

Cottage cheese dip: Blend one pound of homemade cottage cheese with a teaspoon each of ground celery seeds, dill seeds, and caraway seeds. Add a tablespoon each of minced parsley and chives. When well-blended, chill, and allow the flavor to permeate the dip. Dust the top lightly with sweet paprika.

German-style koch kaese: Blend together a quart of homemade cottage cheese with one teaspoon of sea salt and three tablespoons of whole caraway seeds.

Cover the bowl, and set it in a warm place. Stir the mixture daily with a fork, for a week's time, or until the cheese is "ripe" and clear. Heat one tablespoon of oil and 1½ cups of water. Add the cheese. Simmer for twenty minutes, stirring constantly. Then remove the pot from the heat and cool the mixture. When cool, blend in one beaten egg yolk, and continue to beat until the mixture is glossy. Chill the cheese. Serves six.

Cottage cheese dressing: Blend together in an electric blender ½ cup homemade kefir, ½ cup homemade cottage cheese, ½ cup apple cider vinegar, ¼ teaspoon sea salt, three hard-cooked egg yolks and ½ green pepper until smooth, and pour over wedges of lettuce. Makes 1½ cups of dressing.

Quickie cottage cheese dessert: Blend together in a bowl two cups of dry homemade cottage cheese, ¼ cup raisins, three tablespoons of honey, and ¼ teaspoon ground cinnamon. Pile the mixture into sherbet glasses, and garnish with chopped nuts and fresh fruit. Serves six.

Russian Easter dessert: Blend together in a bowl ¾ pound of dry homemade cottage cheese with ½ cup of homemade kefir, ½ cup of softened butter, one cup of ground nuts, and one cup of dried, chopped fruits. Line a mold with a piece of cheesecloth, and press the mixture into the mold. Place a weight on top to press it down. Let it stand overnight. Unmold carefully the next day. Serves six to eight.

Dutch cheese spread: Pour one quart of fresh sweet milk into an earthen bowl, and allow it to stand in a warm place until it thickens. Then pour boiling water over it. Place it in a cheesecloth bag and let it drain for twelve hours. Rub the cheese through a fine sieve. With a spoon, work in two tablespoons of milk and two tablespoons of yogurt until the mixture is the consistency of apple butter. Season with ¼ teaspoon of sea salt. Spread this mixture on buttered bread, and eat with apple butter. Makes about one cup.

Cup cheese: Using a long, sharp knife, cut through one gallon of thick, sour milk several times. Then heat the milk slowly to 90° F. or scald until the curd is very dry. Remove from heat and place it in a wet cheesecloth bag. Press under a heavy weight from twelve to twenty-four hours, or until the cheese is dry. Force the cheese through a sieve, or grate it fine. Place it in a wooden bowl, cover it with a heavy cloth, and keep it in a warm place from three to seven days, or until it is soft and ripe. During this time, stir it occasionally. Then place it in a skillet, and cook it over a low heat, stirring constantly, until all of the lumps are dissolved. Add one teaspoon of sea salt and three tablespoons of butter. Mix well and turn the mixture into cups. Chill. Makes three cups.

◆ ◆ ◆

Homemade hard cheese: If you have access to raw cow's or goat's milk, you can make good homemade hard cheese.(6) Allow one gallon of clean, raw evening milk to ripen overnight in a cool place, from 50° F. to 60° F. The following morning, add to it a gallon of clean, raw morning milk. This mixture will yield a better cheese than if all fresh milk is used. Above all, the milk must be clean and taste sweet.

Warm the milk to 86° F. in a clean, sound, enameled or stainless steel pot. Dissolve ¼ of a cheese rennet tablet* in a glass of cold water. To help dissolve the tablet readily, break and crush it with a spoon. Stir until it is completely dissolved.

---

*In this instance, the ¼ of a rennet tablet is Hansen's; if you use Junket rennet, use three tablets. (See footnote on page 90.)

Set the pot of milk in a larger vessel of warm water, 88° to 90° F. in a warm place away from drafts. Add the rennet solution. Stir the milk thoroughly.

Let the mixture stand *undisturbed* until a firm curd forms—within thirty to forty-five minutes. Test the firmness of the curd by carefully putting your finger into the curd at an angle and lifting it. If the curd breaks clean over your finger, it is ready to cut. If it does not break clean, allow it to set longer.

Remove the pot of curd from the larger vessel. Cut the curd into small cubes, using a long butcher knife or spatula. The instrument should be long enough so that the blade will go clear to the bottom of the pot without the handle dipping into the curd. Cut into squares of about ⅜ inch. Then, using the knife at an angle, make additional cuts. Then turn the pot around and cut similar angular cuts in the opposite direction.

Using your hand, stir the curd very gently but thoroughly, using long, slow movements around the pot, and from the bottom up. Carefully cut up the larger pieces that come up from the bottom, but do not squash the curd. Try to make the pieces of curd as nearly the same size as possible. Stir continuously by hand for fifteen minutes, to keep the curds from sticking together.

Then place the pot in the larger vessel of hot water, as a double boiler. Heat the curd slowly by raising the temperature of the curd and whey about 1½° every five minutes until it reaches 102° F. Stir it with a spoon occasionally to keep the curds from sticking together. Heating should continue slowly. If necessary, heat a few degrees above 102° F., or until the curds hold their shape and readily fall apart when held on your hand without being squeezed.

Remove the pot from the heat. Stir it every five to ten minutes to keep the curds from matting together.

Leave the curds in the warm whey until the mass becomes firm enough so that the pieces, when pressed together in a handful, will easily shake apart. This will take one hour.

Turn the curds into a three to four foot square of clean cheesecloth. Holding two corners of the square cloth in each hand, swing the cloth gently to allow the curds to roll back and forth without sticking together for two to three minutes. During this time, the whey will drip through the cheesecloth.

Place the cheesecloth with the curds in an empty clean pot. Sprinkle one tablespoon of sea salt over the curds. Mix well by hand, but do not squeeze. Then add an additional tablespoon of sea salt and mix well again.

Tie the four corners of the cheesecloth crosswise, forming the curds into a ball. Hang this cheesecloth bag up, and allow the whey to drip through for ½ to ¾ of an hour.

Remove the cheesecloth from the ball of curds. Fold a long, clean cloth, shaped like a dish towel, into a bandage about three inches wide, and wrap it tightly around the ball. Form it into a flat, round shape. Pin the cloth into place. With your hand, press down the curds to make the top surface of the cheese smooth. There should be no cracks extending into the center of the cheese. The round should be not more than six inches across, to prevent too much drying out.

Place three or four layers of cheesecloth under and on top of the cheese. Place the round on a cheese press. (This is simply constructed. It consists of two pieces of wood, eight by twelve inches, through which two one-inch dowels are driven, one on each side.) The cheese is placed on the lower board, and the upper board is placed on top of the cheese. The dowels help keep the boards in place. Place two bricks on top of the upper board, to press down the cheese. At night, turn the cheese over, return it to the cheese press, and place four bricks on top. Allow the cheese to stand until morning.

Remove the cloths from the cheese, and place the round on a board for half a day. Turn it occasionally, to help the rind become completely dry. Then dip the round in hot liquid paraffin. Dip first one half, hold a minute, then dip the other half. Or, if you prefer, paint on the liquid paraffin with a brush. Or rub vegetable oil on the cheese. Store in a clean, cool but frost-free cellar, or similar place. Each day, for a few days, turn over the cheese round. Then, turn it two or three times weekly. Usually the cheese is sufficiently ripened to eat within three to four weeks.

A homemade grating-type Italian cheese can be made as follows: Heat three gallons of fresh milk to 85° F. and add a solution of twelve junket rennet tablets or one Hansen's cheese rennet tablet dissolved in a little cold water. (See footnote on page 90.) Allow the mixture to stand until the curd is quite firm. This should take about forty minutes. Break up the curd with your hands. Then heat the mass in the whey until it is as hot as your hand can tolerate. Then gather the curds by hand, and press the mass together until it is firm.

Drain off the whey, and place the curds in a cheese hoop. Press until the mass shapes itself well in the hoop, then reverse it in the hoop, and continue to press until it is very firm. Put the hoop, containing the curds, back into the whey and heat it carefully until *just below* boiling, but *not boiling*. Remove from the heat, and allow to cool in the whey.* When cold, remove from the whey and allow to drain for twenty-

*Retain this whey for making homemade Ricotta cheese. (See pages 92-93.)

four hours. Then remove it from the hoop. The cheese, soft and quite palatable, can be eaten at this stage. However, to prepare it as a grating cheese, rub it with salt and allow it to rest on a shelf in a cool temperature for three to four days, or until it is quite dry. Then place it in a jar of strong brine for another three or four days. Remove it from the brine, wipe it dry, and place it once again on a shelf in a cool place to dry for a few more days. Place the cheese in a crock and allow it to cure from four to six months. During the first month of curing, rub it lightly with salt once a week, or until the rind is dry. Then cover the cheese with a cheesecloth, and place it in the crock.

In making hard cheeses, if you want a product to be of a harder consistency, cut the curd into smaller pieces. Also, heat the milk a little longer, or raise the temperature slightly in heating. Use more weight in pressing.

If you want a softer cheese, cut the curd into somewhat larger pieces. Also, heat the milk for a shorter period, or lower the temperature slightly in heating. Use less weight in pressing.

One type of hard cheese is made from whey, the by-product of cheesemaking or buttermaking. The famous goat cheese of Norway is made from whey. In the kind of storage economy that dominated the Scandinavian households in past centuries, nothing edible could be wasted. Out of this sense of thrift whey cheese was devised. Since whey is little more than water, its boiling down took hours to complete. It formed a brown paste that was put into molds and allowed to set. (For more information about whey, see pages 113-115.)

Five-minute mock soufflés: Blend together in an electric blender six eggs and ½ cup of sharp Cheddar cheese until the mixture is smooth. Turn it into buttered poaching cups, and cook gently until firm. Serves four to six.

Cheese and onion pie: Blend together in an electric blender one scant cup of milk, three eggs, one tablespoon of vegetable oil, one teaspoon of arrowroot flour, ¾ cup of sharp Cheddar cheese and ⅛ teaspoon of chervil. Blend until smooth. Turn the mixture into a buttered pie plate. Dice two medium-size onions, and scatter them through the mixture. Top with two tablespoons of sesame seeds. Bake in a preheated oven at 300° F. for one hour. Serve hot or cold, with sweet cider. One pie serves two persons generously. To double the recipe, simply repeat the same operation in the electric blender, and cut additional onions.

Cheese-cauliflower soup: Cook a medium-size head of cauliflower (about three cups), and break it into small cauliflowerettes. Allow them to cool. Sauté a small, finely chopped onion in two tablespoons of safflower oil until transparent but not brown. Allow to cool. In an electric blender combine one pint of cold buttermilk with one tablespoon of arrowroot flour and one cup of cubes of sharp Cheddar cheese. When the mixture is smooth, gradually add the cauliflower and onion. Turn the mixture into the top of a double boiler. Add an additional pint of cold buttermilk. Gently heat the mixture until it is hot, stirring during the heating time. Remove the pot from the heat and serve immediately. Garnish with minced parsley. Serves six.

◆ ◆ ◆

Fried-milk curd, a newly developed fermented milk product: Scientists of the U.S. Department of Agriculture have developed a food that would make both Miss Muffet and the spider sit up and take notice. It is a deep-fried milk curd. To make the new food, chemists add calcium chloride to skim milk to form the curd. (This substance is frequently used in processing milk to produce curd.) The skim-milk curd is heated, cut into bite-sized pieces, and deep-fried in hot oil until slightly brown. The resulting product is meatlike in texture, does not fall apart in prolonged heating, is bland, and has good storage qualities. The texture as well as the flavor can be modified to suit individual tastes. Fried curd can be canned in a meat-flavored gravy, or it can be used for snacks, hors d'oeuvres or confections. Soaked in water, the fried curd will keep under refrigeration for two weeks; at room temperature, sterilized curd will keep three months. The new food is high in nutritious milk protein, and obviously it is quite different from the traditional forms of milk and dairy products. People who reject milk may find fried milk curds acceptable as a dairy food in their diets. From a nutritional viewpoint, it would be desirable to experiment further with milk-curd processing to learn whether a satisfactory product could be made by baking or broiling, instead of by deep-fat frying.(7)

Notes

1. Vivienne Marquis and Patricia Haskell, *The Cheese Book,* pp. 17-29. New York: Simon & Schuster, Inc. 1964-1965.
2. *Ibid.,* p. 20.
3. "Cheese," *Encyclopedia Britannica,* 1948, Vol. 5, pp. 333-336.
4. Vivienne Marquis and Patricia Haskell, *Op. cit.,* pp. 33-44.
5. "Ultra-Pasteurized Cream Has Fine Flavor, Long Shelf-Life," *Yankee Dairy News,* July 1972, Vol. 1, No. 1.
6. Clement A. Phillips, *Making Cheese at Home.* One-

Sheet Answers No. 90, Dept. of Food Science and Technology, Agricultural Extension Service, University of California, 1963; H. E. Walter, *An American-Type Cheese; How to Make It for Home Use.* USDA Farmers' Bulletin No. 2075, 1962; *How to Make Cheese.* Illustrated wall chart, C. Hansen's Laboratories, Inc., Milwaukee, Wisconsin 53214, no date.
7. "How Do You Like Your Milk?" *Service USDA's Report to Consumers.* Feb. 1971, p. 2.

SOUR CREAM, BUTTERMILK & WHEY
Fermented Milk Products

Homemade sour cream: Sour cream can easily be prepared at home.(1) All that is needed is a bottle of sweet cream, some buttermilk, and a teaspoon.

Generally, the heavier the sweet cream (that is, the higher the percentage of butterfat) the better the body of the sour cream. However, a very satisfactory product can be made from coffee cream—cream containing about 20 percent of butterfat.

To obtain the best quality and flavor in the sour cream, make certain that the original sweet cream is of good quality and flavor. Also, it is preferable to use sweet cream that has been pasteurized, so that no undesirable bacteria will be present to develop and spoil the flavor of the sour cream.

If commercially pasteurized cream is unavailable, cream may be pasteurized at home in a double boiler. Use a double boiler that holds twice as much cream as you wish to pasteurize. For example, a quart-size pot should be selected for a pint of cream, or a gallon-size pot for two quarts of cream. Pour the chilled cream into the lower compartment. Have the cold water touch the bottom of the upper vessel so that the cream will begin to heat as soon as the water does. Keep a thermometer in the cream. Bring the temperature of the cream to between 155° and 160° F. Hold it there for thirty minutes. After the thirty-minute period is finished, cool the cream rapidly by removing the top vessel that contains the cream, and set this pot in cold water, preferably ice water. Change the water when it becomes warm. If the buttermilk is to be added immediately, the cooling process should be stopped when the cream reaches a temperature of 85° F; otherwise, the cream should be cooled to below 50° F. and placed in a refrigerator immediately and kept there until it is to be used.

The object of adding buttermilk is to introduce a large number of active milk-souring bacteria into the cream. When large numbers of these active milk-souring bacteria are present, they not only sour the cream quickly by growing rapidly, but they play a vital role in preventing the development of other bacteria that might give the cream undesirable flavors and odors. It is important, therefore, to use high-acid buttermilk with a clean flavor.

Most dairies sell what is known as "cultured" buttermilk. (See pages 111-112.) This is made by souring clean, pasteurized, skimmed or partly skimmed milk. Commercial buttermilk usually has about 1 percent acid. Buttermilk from farm churning may contain very little acid, or it may have considerably more than commercial buttermilk. Farm-churned buttermilk may contain undesirable types of bacteria, as the cream is not usually pasteurized or ripened under controlled conditions. All factors considered, cultured buttermilk is usually the preferable starter for making homemade sour cream.

Having both the bottle of prepared cream and the bottle of buttermilk on hand, the next step is to add the proper amount of starter. After thoroughly mixing the bottle of cream, pour out about half of it into a thoroughly cleaned container. Shake the bottle of buttermilk well, and for each pint of cream to be soured, put five teaspoonfuls of buttermilk into the measured cream. Pour back enough of this mixture into the cream bottle to within three-fourths of an inch from the top. Stopper this bottle and shake it until the contents are thoroughly mixed. Then set the bottle of cream in a warm place (70° to 85° F.) for twenty-four hours. Although the sour cream may then be used, its body will be improved if it is refrigerated for another day.

In cold weather, a larger amount of buttermilk may be added to hasten the souring of the cream. However, a better method is to hold the souring cream at room temperature for an extra twelve to twenty-four hours.

Some sour-cream dishes require that the cream be whipped. When sour cream is not to be cooked, it can be made smoother and stiffer by whipping. Care must be taken, however, not to overwhip or it will turn into butter. To avoid this, and to hasten whipping, keep the cream cold while whipping.

(For nutrients in sour cream, see Tables I and II, pages 36-39.)

Sour cream can be used extensively in the preparation of foods, especially as an ingredient in dressings, as a garnish, and as an ingredient in baking and cooking. It can be used interchangeably in many recipes for buttermilk or yogurt. It has more body than either buttermilk or yogurt. Sour cream can be used in quick breads, cakes and cookies. Because one cup of heavy sour cream contains approximately 40 percent fat (about six tablespoons), sour cream can be used to replace part or all of the fat called for in recipes for pancakes, waffles, muffins, biscuits, cakes and cookies, as well as to replace the milk. In batters calling for a large amount of liquid, the heavy sour cream may contain more fat than the amount called for in the original recipe, and a richer product will result.

Sour cream-cucumber sauce: Blend together in an electric blender one cup of homemade sour cream and two small raw cubed cucumbers. Chill, and serve over hot or cold fish. Makes three cups of sauce.

Sour cream sauce for vegetables: Blend together in an electric blender one cup of sour cream and a half cup of fresh parsley, chives and green pepper. Blend thoroughly, and use over asparagus, broccoli, cauliflower, spinach, brussel sprouts, or baked potato.

Sour cream-potato soup: Cook two cups of cubed raw potatoes in three cups of seasoned soup stock, with one onion, sliced fine, until the potatoes are soft. Remove from the heat and fold in a cup of homemade sour cream. Serve at once, topped with fresh minced parsley. Serves six.

Sour cream dressing: Blend together in an electric blender one cup of homemade sour cream, one tablespoon of honey, ¼ teaspoon of sea salt, three tablespoons of safflower oil, and two tablespoons of apple cider vinegar. Serve over a tossed green salad, hearts of lettuce, cucumbers, or cole slaw. Makes 1⅓ cups of dressing.

Sour cream sherbet: Soak two cups of prunes in warm water, drain and remove the pits. Blend them together in an electric blender with one cup of homemade sour cream and ⅛ teaspoon of freshly grated nutmeg. Turn into sherbet glasses and chill. Serves six.

◆ ◆ ◆

There are two types of buttermilk, and they are distinctly different products. The first is real churned buttermilk, a by-product of butter making. The second is cultured buttermilk, the type commonly sold commercially.

Churned buttermilk is the residue of cream after the butter has been removed by churning. It consists mainly of water, and has some milk sugar, casein, butterfat and lactic acid. It has a slightly acid taste due to the lactic acid, which is formed during the ripening of the cream.

The lactic acid and acidophilus content of farm-produced buttermilk is variable. Until about 1900, American farmers fed most of the buttermilk to their pigs. In the countryside of Ireland and Scotland it was used by the farm families on their potatoes and with their porridge. A suggestion was made in Great Britain that the buttermilk could make a valuable nutritional contribution if it were sold at a low price to the poor people in the city slums.(2) Gradually, the healthful and nutritious qualities of buttermilk were recognized (see page 18), and it became especially favored as a refreshing hot-weather drink. It is light and easily digested.

Cultured buttermilk is a process developed as a response to the growing consumer interest in this beverage. It is made commercially by fermenting skim milk. A vat is filled in the evening with skim milk and brought to a pasteurizing temperature. Then it is cooled to 100° to 105° F. A starter or culture is added, the batch is stirred, and it is maintained at a desired temperature until the skim milk has fermented sufficiently to develop an acidity of at least 0.6 percent lactic acid. By then, the batch is also filled with colonies of acidophilus bacteria. The partially fermented skim milk is poured into churns, some butter is added, the mass is churned and the butter is broken up into spicules to *simulate* farm buttermilk. Since the percentage of both lactic acid and acidophilus bacteria can be controlled in cultured buttermilk, its nutritional qualities are less variable than those in farm-churned buttermilk, and it may be superior.(3)

Buttermilk has, in addition to whatever nutrients are present in the original milk, the valuable lactic acid. This substance helps to boost the degree of buffered acidity of human gastric juices, which many people lack in sufficient quantity.

Buttermilk quencher: Blend together in an electric blender one quart of buttermilk and one pint of unsweetened pineapple juice. Chill, and serve with sprigs of mint. Serves six.

Buttermilk smoothie: Blend together in the electric blender one quart of buttermilk, two ripe bananas and two tablespoons of unsulfured molasses. Serves six.

Buttermilk nog: Blend together in an electric blender one quart of buttermilk and one pint of unsweetened orange juice. Add ½ cup of wheat germ, and blend thoroughly. Chill. Serves six.

Quickie buttermilk dessert: Blend together in an electric blender one pint of buttermilk with two cups of soaked, drained dried apricots. Blend thoroughly until smooth. Turn into bowls. Chill, and garnish with ground nuts. Serves six.

Summer beet soup: Cook one cup of beets and allow them to cool. Peel and dice. Place one quart of seasoned stock in an electric blender, add the beets, ½ teaspoon of sea salt, a sprig of parsley, a sprig of dill, and a pint of buttermilk. Blend thoroughly. Chill before serving. Serves six.

◆ ◆ ◆

Whey is a by-product of fermented milk foods such as cheese or yogurt. When you make these foods at home, by all means plan to use the whey. When you drain the whey from the cheese curds, place a basin under the hanging sack to catch the dripping whey. If you make yogurt, carefully drain off any liquid whey that may form on top, and make use of it. Whey retains all of the water-soluble vitamins, and the lactose of the milk. Either drink the whey neat, or add it to such things as soup stock, or use it instead of some of the water in bread dough. According to old Norwegian farmers who used to drink whey in the fields, no more thirst-quenching drink could be found in the summertime.

Whey was long considered a problem by the dairy industry, for cheesemakers had only been able to use limited amounts of this by-product. A hundred pounds of milk will yield about ninety pounds of whey and ten pounds of cheese. Only about a third of the twenty-two billion pounds of whey produced yearly in the United States from cheese processing went into food use. Some was used to feed hogs, and some was used to fertilize fields. On occasions when the price of dry milk was high, whey was sold as a dry skim milk substance. The remainder of the whey was frequently poured into rivers, which created a serious waste disposal problem of pollution.

With the severe food shortages in developing countries, the use of low-cost, highly-nutritious dried whey has been viewed as a valuable addition to United

States aid programs. The market is being further increased by the addition of whey to specialty foods. It improves the nutritional value of foods and also adds other desirable characteristics. Dried whey in bread, for example, gives a golden color when the bread is toasted. For this, and similar reasons, dried whey is now being used in hot roll mixes, pop-tarts, frozen baked potatoes or frozen vegetables with cheese sauce, sour cream sauce, candies and cookies, frozen cream pies, instant western omelets, frozen fish sticks, and other convenience foods.(4)

Two diverse reports in recent years offer promising uses for whey as a nutritional aid and as a therapeutic one. British researchers have been investigating the use of dried whey in ice pops—a commodity high in sugar and acid. The inclusion of only 2.6 percent whey in the mix reduced the acidity of the ice pops to a level that was found effective in reducing tooth decay caused by the consumption of water ices. Thus, the addition of whey not only makes the ice pops far more nutritious but may alleviate tooth decay associated with the consumption of acidic and sugary foods.(5)

Uremia is a condition marked by the presence of urinary constituents in the blood, accompanied by a toxic condition. Patients with uremia may be severely limited in the choice of foods due to stringent dietary restrictions. To lessen the burden of adhering to a usually very monotonous diet, a researcher developed products resembling ordinary foods, using electrodialyzed whey (whey so treated that substances in solution are separated; the noncolloids are separated from the colloids, the protein) as the sole source of protein. Recipes were developed for beverages, cheese-type spreads and desserts, using the electrodialyzed whey powder as the basic ingredient.(6)

According to U.S. Department of Agriculture information, whey has the following nutrients, based on 100 grams (the first set of figures are for fluid whey; the second set, for dried whey): food energy, 118 and 1,583 calories; protein, 4.1 and 58.5 grams; fat, 1.4 and 5.0 grams; carbohydrates, 23.1 and 333.4 grams; calcium, 231 and 2,930 milligrams; phosphorus, 240 and 2,672 milligrams; iron, 0.5 and 6.4 milligrams; vitamin A, 50 and 230 I.U.; thiamin, 0.15 and 2.26 milligrams; riboflavin, 0.65 and 11.39 milligrams; and niacin, 0.3 and 3.6 milligrams.(7)

(For a description of Norwegian whey cheese, see page 98; for a discussion of the bactericidal properties of whey, see page 10.)

A lactic-acid fermented whey product may be prepared with *L. acidophilus* on which animals thrive. When fed this whey product, the health of the animals improved; they became less nervous, more easily controlled, and had larger weight gains.(8)

Whey has potentialities as a useful food preservative. Presently, whey-fermenting bacteria are being added to cottage cheese to extend its shelf life. A small heat-stable compound, a peptide, has been isolated from the bacterial fermentation of dried whey which inhibits the bacteria that cause spoilage in refrigerated fresh meat, fish and milk. Even more important, this compound inhibits certain pathogenic organisms, including *Pseudomonas aeruginosa*.(9)

NOTES

1. *Sour Cream: How to Prepare and Use It At Home.* Leaflet No. 213, USDA, Oct. 1941.

2. "Buttermilk," *Chamber's Encyclopedia*, 1895, Vol. 2, pp. 585-586.

3. *Modern Nutrition*, Aug. 1963, p. 10.

4. "Study to Cut Waste Whey." *Service, USDA's Report to Consumers*, May 1971, p. 3.
5. "Whey: Now An Asset." *Agricultural Research*, USDA, Aug. 1967, p. 7.
6. N. R. J. Karp, "Electrodialyzed Whey-Based Foods for Use in Chronic Uremia." *Journal of American Dietetic Association*, 1971, Vol. 59, pp. 568-571.
7. *Composition of Foods*, USDA, Agricultural Handbook No. 8, 1963, revised, p. 121.
8. "Cultured whey product." Herbert R. Peer, Ferma Gro Corp., U.S. Patent 3,497,359, Feb. 24, 1970; abstracted in *Chemical Abstracts*, June 8, 1970, Vol. 72, No. 12.
9. *Chemical and Engineering News*, Nov. 20, 1972, p. 14.

INDEX